# PROTECTED AREAS AND REGIONAL DEVELOPMENT IN EUROPE

T0248601

# Protected Areas and Regional Development in Europe

## Towards a New Model for the 21st Century

*Edited by*

INGO MOSE
*Carl von Ossietzky University of Oldenburg, Germany*

Routledge
Taylor & Francis Group

LONDON AND NEW YORK

First published 2007 by Ashgate Publishing

2 Park Square, Milton Park, Abingdon, Oxon OX14 4RN
711 Third Avenue, New York, NY 10017, USA

*Routledge is an imprint of the Taylor & Francis Group, an informa business*

First issued in paperback 2016

**British Library Cataloguing in Publication Data**
Protected areas and regional development in Europe :
   towards a new model for the 21st century. - (Ashgate
   studies in environmental policy and practice)
   1. National parks and reserves - Europe - Case studies
   2. Protected areas - Europe - Case studies 3. Regional
   planning - Europe - Case studies
   I. Mose, Ingo
   333.7'8316'094

**Library of Congress Cataloging-in-Publication Data**
Protected areas and regional development in Europe : towards a new model for the 21st century / edited by Ingo Mose.
      p. cm. -- (Ashgate studies in environmental policy and practice)
   Includes index.
   ISBN 978-0-7546-4801-7
   1. Protected areas--Europe. 2. Regional planning--Europe. 3. Nature conservation--Europe. 4. Sustainable development--Europe. I. Mose, Ingo.

   S934.E85P76 2007
   333.72094--dc22

                                                                    2006039321

ISBN 13: 978-0-7546-4801-7 (hbk)
ISBN 13: 978-1-138-26262-1 (pbk)

# Contents

**PART III      Synthesis**

# List of Figures

# List of Tables

# List of Contributors

**Marco Aufdereggen**, University of Applied Sciences, Research Center for Leisure, Tourism and Landscape, Institute of Landscape and Open Space, Rapperswil, Switzerland. Marco is a Swiss jurist who received an LL.M in Environmental and Natural Resources Law from Lewis and Clark College in Portland, Oregon. He is particularly interested in the field of energy law.

**Christine Gamper**, M.A., is a scientific assistent at the Faculty of Interdisciplinary Studies (Department of Urban and Regional Studies, Vienna), University of Klagenfurt, Austria. Her key aspects of activity are cultural landscape research, regional/urban development and spatial planning.

**Thomas Hammer**, Ph.D., is a Senior Lecturer at the Interfacultary Coordination Office for General Ecology (IKAÖ) at the University of Berne and at the Institute for Geography at the University of Fribourg, Switzerland. He manages several inter- and transdisciplinary research projects focussing on the link between sustainable landscapes and regional development inside and outside of protected areas. His second field of research interest concerns the link between desertification and regional development in the sahelian countries of West Africa.

**Martin Heintel**, Ph.D., is an Associate Professor in Human Geography at the Department of Geography and Regional Research, University of Vienna, Austria. His main areas of research are spatial development, comparative regional development, enlargement of the European Union (with special emphasis on cross-border cooperations), and third world mega-cities. He is actually a Visiting Professor at the University of New Orleans (Marshall Plan Chair), USA, and has conducted further visiting lectureships at BabeÅŸ-Bolyai University Cluj-Napoca, Romania, Humboldt-University Berlin, Germany, and University of Salzburg, Austria.

**Stefan Kah**, Dipl.-Geogr., works at ORMO, the regional development office of the Moesano region in the Italian speaking part of the Swiss Canton Graubuenden, where he is currently involved in the conceptual planning process for the future national park 'Parc Adula'. He is also contributing to various projects, including Interreg IIIA activities. His other interests include economic and tourist development, especially on a European scale.

**Markus Leibenath**, Ph.D., studied Landscape Ecology and Resource Economics at Munich Technical University, Germany and at Harvard University, USA. He earned his doctoral degree at Berlin Technical University with a thesis on the marketing of national park regions. Since 2000 he has been working as research associate at the Leibniz Institute of Ecological and Regional Development in Dresden, Germany. Recently he has published on transboundary spatial development, on European biodiversity policies and on economic evaluation of protected areas. He is also teaching at Dresden Technical University and co-ordinates the research network spa-ce.net.

**Michael Leitner**, Ph.D., is an Assistant Professor in the Department of Geography and Anthropology at Louisiana State University in Baton Rouge, USA. He received his B.A. and M.A. in Geography from the University of Vienna, Austria and his Ph.D. in Geography from the State University of New York in Buffalo, USA. His main research and teaching interests are in cartographic visualization, the research and application of Geographic Information Science (GISc) to public health, public safety and forensic analysis. His research is highly interdisciplinary and overlaps primarily with Forensic Anthropology, Sociology, Criminology and Criminal Justice, Statistics, Public Safety, and Psychology.

**Florian Lintzmeyer**, Dipl. Geogr., is Research Assistant at the Chair of Land Readjustment and Rural Development at the Technical University of Munich and freelancer for ifuplan, Institute for Environmental Planning, Munich. He is currently involved in the INTERREG IIIb-project DIAMONT ('Data Infrastructure in the Alps: Mountain Oriented Network Technology'), conducting research on indicators for spatial trends in the Alpine region. His further interests include nature-based tourism and its regional economic effects and the potential of protected areas for regional development in the Alpine context.

**L. Rory MacLellan** is currently Senior Lecturer in Tourism at the University of Strathclyde in Glasgow, Scotland, UK. His teaching and research interests include public sector tourism organisations, rural tourism, sustainable tourism and tourism and the natural environment. He has published on aspects of tourism and the natural environment of Scotland including articles on wildlife holidays, national parks, sustainable tourism and has co-edited a book titled 'Tourism in Scotland'. He has carried out research and consultancy on aspects of Scottish tourism for a number of public agencies and remains actively involved in managing tourism and the natural environment of Scotland.

**Isabelle Mauz**, Ph.D., is researcher in Sociology and member of the research unit "Development of mountain regions", Cemagref of Grenoble, France. She is involved in research regarding human relationships to nature and has particularly studied the history and the memories of the French protected areas, as well as the role of wild animals in the social construction.

**Ingo Mose**, Ph.D., is a Geographer and Regional Scientist holding a professorship at the Carl von Ossietzky University of Oldenburg, Institute of Biology and Environmental Sciences (IBU), Regional Sciences Working Group, Germany at present. His main areas of research comprise regional policy and regional governance, rural development, protected areas, and tourism. He has carried out empirical research in several European countries, including Austria, the UK and Sweden, with major emphasis on qualitative methods. He has been a visiting professor to a number of foreign universities, such as the Universities of Vienna and Salzburg, Austria, Keele University, UK and Södertörns University College, Sweden. Among his publications are numerous articles related to the subject of the book project. He is a member of the *Deutsche Akademie für Landeskunde* and functions as a speaker of the Working Group of Rural Geographers in Germany.

**Birgit Nolte**, Ph.D., is an independent free-lance expert for international and national issues regarding tourism in nature, located in Wismar, Germany. She wrote her doctoral thesis in human geography about 'Tourism in Biosphere Reserves in East Central Europe. Expectations, Obstacles and Opportunities for the Implementation of Sustainability'. Her further interests cover the geography of tourism in protected areas as well as the development of rural areas in Europe.

**Jarkko Saarinen**, Ph.D., is Professor of Human Geography at the University of Oulu, Finland. He was formerly a Professor of Nature-Based Tourism at the University of Lapland and Director of the Pyhätunturi National Park, Finnish Lapland. His research interests include tourism development and its socio-cultural and economic impacts, sustainability and the construction of local identities, culture and nature in tourism. He is currently a chairperson of the International Geographical Union's Commission on Tourism, Leisure and Global Change. He is also Docent in the Finnish University Network for Tourism Studies.

**Dominik Siegrist**, Ph.D., is chair of the Research Centre for Leisure, Tourism and Landscape at the University of Applied Sciences Rapperswil, Switzerland. His main topics are nature based tourism, visitor management, regional development and parks. He is president of the International Commission for the Protection of the Alps (CIPRA).

**Harry Spiess** is lecturer in Economic Geography and researcher at the Institute for Sustainable Development at Zurich University of applied science, Winterthur, Switzerland. His interests cover the field of sustainable location development and sustainable mobility. He is actually involved in research optimising planning instruments for Local Recreational Areas and in a European project to develop Scenarios for Regional Economic and Research Policies. He has published a scenario-approach for value added and subsidies in the Swiss Alpine region for 2030.

**Andreas Voth**, Ph.D., is an Associate Professor in Geography at the University of Vechta, Germany. He is actually continuing research in rural development in the EU, especially in Spain. His main interests in Human Geography cover economic geography, tourism and regional development. He has published widely about agriculture and rural areas in Southern Europe and Brazil.

**Norbert Weixlbaumer**, Ph.D., Associate Professor, University of Vienna, Institute of Geography and Regional Research, Austria. At present he is involved in research regarding the direct and indirect economic benefits of Nature Parks in Austria. His further interests cover rural geography and regional development. From 2000–2006 he was president of the International Commission for the Protection of the Alps (CIPRA), Section Austria.

# Foreword

Ingo Mose

Protected areas policies in Europe are a subject of controversial discussion. This applies to the political debate as well as to the academic and is also well reflected by controversial arguments among the public regarding the implementation of protected areas. In addition to other aspects, the present discussion focuses primarily on the objectives of protected areas policies and their implementation via different types of protected areas.

Protected areas have experienced considerable change with regard to their objectives and tasks. While originally meant to function as reserves for beautiful landscapes and endangered species, they later became connected with the idea of preserving whole ecosystems and their dynamics of natural processes relatively free from human impacts. Today, many protected areas are undergoing change again: Especially large protected areas are increasingly considered to function also as instruments for regional development. This perspective applies particularly to many peripheral rural areas throughout Europe that are faced with severe problems due to economic and socio-cultural disparities. Expectations are high that protected areas could be used as laboratories for experimental projects or even as models for sustainable regional development, based on the endogenous resources and potentials of the region and their development via a specific protected areas policy.

According to different sources, this new understanding of protected areas has been described as a significant shift of paradigm regarding the underlying concept of nature conservation. In rather sharp contrast to traditional concepts, focusing mainly on the conservation objective, and often trying to restrict human activities in protected areas to a very high degree, new approaches are aiming at a consistent integration of conservation and development functions making protected areas real 'living landscapes'. Agriculture as well as forestry, handicrafts, tourism or education offer potential arenas to test in which way and to what extent this process of integration could be developed in practice.

As a result of this discussion, the last decade has seen a noticeable increase of modern and innovative types of protected areas throughout Europe, such as Biosphere Reserves or Regional Parks, which are expected to be models for a new dynamic interpretation of nature conservation. At the same time, even some traditional protected areas are also undergoing significant change towards a greater integration of conservation and development functions. This shift of paradigm is regarded as a major strategic and instrumental challenge for both nature conservation and regional development, resulting in a number of various new tasks for regional planning as well as regional policy.

Experiences with new concepts of protected areas have been very different throughout Europe so far. While in some countries new approaches have been a subject of continuous discussion and empirical testing, only little or hardly any change can be identified elsewhere. Additionally, the general framework of relevant policies needs appropriate consideration, such as EU regional policy and national traditions of regional planning and regional development characterised by a number of considerable differences. As a result, the present picture of protected areas (still) appears rather diffuse, if not confusing.

Against this background, this anthology seeks to deliver a comprehensive overview regarding the relation of protected areas and regional development policies in Europe, covering both the theoretical and practical dimension of the subject. With regard to the latter, it is the clear intention of the book to present a broad cross selection of case studies from various countries throughout Europe to allow for a substantial comparison of different concepts, strategies and instruments being used. Although there has been a growing number of publications contributing to this subject over the last years, these are rather limited in their concepts: they either discuss the question of protected areas or regional development separately, or they only focus on a very small selection of case studies, many of which have been repeated again and again, without a real comparative European perspective.

The structure of the book is based on three elements:

The first part of the book covers main theoretical aspects relevant to the subject. Firstly, it gives an overview of the historical development of nature conservation and protected areas from the early beginnings in the 19th century up to the present, including a comparison of different approaches being used, which serves as a basis for a critical discussion of the recent paradigm shift in protected areas policies and the underlying philosophies of protection and development in Europe. Secondly, against the outlined background a further contribution focusses on the relation between protected areas and regional development more specifically, explaining different ideas of 'development via protection', and finally discussing conflicts and opportunities of these new approaches within the framework of national as well as European regional development policies.

Making up the core of the book, the second part offers a selection of eleven case studies of different protected areas throughout Europe. The main objective of these contributions is to analyse and discuss the relation of protected areas and regional development from a practical perspective. The presentations cover nearly all relevant types of large protected areas being known and implemented in Europe, including national parks, nature parks (according to German and Austrian conservation law), regional nature parks (an alternative type of park developed in the Romanic speaking countries), and biosphere reserves. Furthermore, each single case study will relate to a different European country, comprising examples from Central Europe (Germany, Austria, Switzerland, France), the UK (Scotland), Scandinavia (Finland), Southern Europe (Italy, Spain) and East Europe (Slovakia). Thus the collection covers both protected areas with a long-standing history, brand-new foundations, and even parks which are currently under designation. To avoid a purely additive approach to the case studies (as is the problem with many previous anthologies), each of them concentrate on analysis and discussion of a number of "core elements" which are (more or less) compulsory for every single chapter. These criteria have been identified by the editor together with the other authors in a discursive process beforehand. These include subjects like the construction of "Leitbilder", relevant forms of regional governance, applied funding systems, planning strategies and instruments, economic success, or areas of conflict, just to name a few.

Finally, as a result, the third part of the book presents a general synthesis of the previous case studies. This is based on a number of comparative questions which have been jointly discussed by a team of four authors out of the group of contributors. Their findings allow to identify and describe similarities and differences, strengths and weaknesses, opportunities and threats. Although the synthesis cannot claim to be representative for all protected areas policies throughout Europe, it offers a very broad overview of the present situation of protected areas and the efforts being made to utilise them for regional development purposes. This finally allows us to relate the synthesis to the theoretical discussion in the first part of the book and to compare the empirical findings with present expectations regarding the possible function of protected areas for regeneration in peripheral rural areas or even for sustainable regional development in general.

The contributors of this book are all experts in the field of protected areas and regional development and have a long-standing reputation for their work either in a national or international context. With the majority of contributors being placed at university institutions, their experiences cover both theoretical and empirical research and have been the subject of continuous publications in national and international journals and anthologies. Finally, the selection of authors covers a broad range of contributions from several countries throughout Europe, allowing for a real cross-European perspective of the book.

This book could not have been made real without the supportive encouragement and practical help of various colleagues and friends. I have gained major motivation and support from intensive discussions with Thomas Hammer, Dominik Siegrist and Norbert Weixlbaumer, who have been among the first supporters of the book project. No question, their input has been of substantial importance for the design of the book concept, and their stimulations have been a major source of my motivation to carry on. However, I would also like to thank all other contributors of the book for their participation in this project. Without their interest and competence, without their flexibility and patience the book might have stranded long before.

Finally, I owe many thanks to my former student Jantje Blatt at the University of Oldenburg. She has spent numerous hours proof-reading the manuscripts and preparing the required lay-out for the publisher. Her interest in the subject and her professional attitude has been indispensable for my work as an editor.

Oldenburg, March 2007

# PART I
# Theoretical Background

Chapter 1

# A New Paradigm for Protected Areas in Europe?

Ingo Mose and Norbert Weixlbaumer

## Introduction

In the course of the global debate on sustainability, protected areas have attained a significant position since, at the latest, the 1990s. According to widely accepted definitions such as the one of the IUCN – The World Conservation Union – the superior objective of protected areas is considered to be nature protection: their tasks are the protection and conservation of biodiversity as well as natural and cultural resources (see EUROPARC/IUCN 1999). Large protected areas have a particularly significant effect. The visible increase in relevance of these so-called large protected areas has less to do with the arbitrarily determined size of at least 1.000 ha, but more with the function increasingly attributed to them as a 'model landscape' for the sustainable advancement of natural and cultural landscapes (see Hammer 2003).

Several international policy papers have helped this notion to find its way into national conceptions of protected areas worldwide. For instance, the Caracas-Conference in 1992 (IV. World Congress on National Parks and Protected Areas) reached an agreement that the advancement of large protected areas should be part of all regional land use planning, namely 'oriented towards sustainable development as well as a "reasonable use" of natural resources' (Revermann and Petermann 2003, 35). At the same time, during this Congress the 'Global Biodiversity Strategy' was adopted, which was succeed by the ratification of the UN 'Convention on Biological Diversity' at the end of 1992. This was the first comprehensive agreement for the protection of the world's biological resources, which aims to designate worldwide at least 12 per cent of the terrestrial surface of the earth as protected areas by the turn of the century. In the meantime, at least on paper, this goal had already been met in Europe with 14.6 per cent designated (see Figure 1.1).

Nevertheless the question arises: does the quantitatively achieved goal of this stage also guarantee that all natural and cultural landscape types are represented and thus adequately protected? Do the established protected areas fulfill expectations as 'model regions' of a sustainable development (see also the critical discussion in McNamee 2002, 59 pp.). The few evaluation-studies that examine the quality control of protected areas thus far (cf. for the situation in Germany and Austria Weixlbaumer 2003, 33) convey a rather doubtful image. In only 10 to 20 per cent of the areas were the goals for protection met. Moreover, the required representation was by no means achieved.

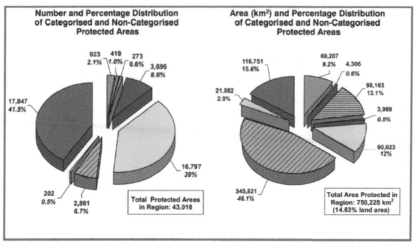

Number and Percentage Distribution of Categorised and Non-Categorised Protected Areas

Area (km²) and Percentage Distribution of Categorised and Non-Categorised Protected Areas

Total Protected Areas in Region: 43,018

Total Area Protected in Region: 750,225 km² (14.63% land area)

Region contains: Albania, Andorra, Austria, Belgium, Bulgaria, Bosnia and Herzegovina, Croatia, Czech Republic, Denmark, Estonia, Faroe Islands, Federal Republic of Germany, Finland, France, Gibraltar, Greece, Hungary, Iceland, Ireland, Italy, Latvia, Liechtenstein, Lithuania, Luxembourg, Macedonia, Malta, Monaco, Netherlands, Norway, Poland, Portugal, Romania, San Marino, Slovakia, Slovenia, Spain, Svalbard and Jan Mayen Islands, Sweden, Switzerland, United Kingdom, Vatican City State (Holy See), Yugoslavia

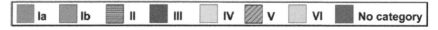

Ia    Ib    II    III    IV    V    VI    No category

**Figure 1.1    Weighting of the IUCN categories of protected areas in Europe**
*Source: Chape et al. 2003, 40*

Despite the fact that the protected areas had only partially met previously established standards, in the 1990s an additional requirement was introduced, which, thus far, still has not been achieved: strengthened implementation of landscape protection outside of the designated protected areas. An example of this can be seen in the campaign of the Council of Europe for 'area-wide nature protection' during the European Nature Conservation Year in 1995.

With the increasing designation of protected areas, despite the significant qualitative deficiencies, at least two important advances have been achieved: a heightened societal awareness of the meaning of area protection and a significant reservation of areas with endangered natural and cultural landscapes important for future area protection policy. During the Fifth World Parks Congress held in Durban in 2003, along with an expression of concern about the current state of protected areas, came substantiation of the enormous success achieved in the last ten years with over 12 per cent of the world's terrestrial surface now designated protected areas.

Large protected areas based on current scientific knowledge are rarely mono-functionally oriented, but rather often fulfill a multitude of different tasks simultaneously. Today the general consensus is that the following goals and functions, which leave the traditional nature protection paradigm behind, are the main focus:

- Preservation and advancement of biodiversity – *regulatory function*
- Regional and supra-regional welfare-effects – *habitat function*
- Gene pool as well as natural disaster-prevention – *support function*
- Sustainable regional development – *development function*
- Environmental education and training – *information function*.

The multi-functional orientation of large protected areas bears in equal measure enormous challenges and substantial conflict-potential. The multitude of diverse expectations for these protected areas has great potential to create conflict about the very purpose of beneficial use of the area. This conflict can break out because of opposing interests between the protected area and its surroundings or even due to clashing interests within a protected area. The spatial overlapping of nature protection and tourism or of nature protection and agriculture exemplifies this issue. Today, it is therefore all the more important to spatially coordinate and steer the diverse goals and functions with all of the different available tools. This is particularly important when it comes to defining and designating different categories of protected areas.

**International and National Categories of Protected Areas**

Throughout the world the role of protected areas is defined by very diverse nationally and regionally protected area concepts. Europe on its own is covered with protected areas of great – oftentimes confusing – diversity. In Germany there are eleven different types of protected areas (see Büchter and Leiner 2000) and in Austria there are twelve (see ÖROK 1997, 18). In order to increase the transparency and at the same time the comparability of protected areas internationally with regards to their goals, the IUCN developed a worldwide category-system of protected areas. Even though this system has gained international recognition, it is not a legally binding tool and is therefore solely voluntary; thus an abolishment of the national categories for protected areas is not intended.

Management objectives and examples for the position of the IUCN-categories become comprehensible by means of Figure 1.1, which in addition illustrates the spatial magnitude of each category in Europe. The strong predominance of Category V (protected landscapes), which takes up by far the largest surface ratio, is striking. Even if, according to the official IUCN-diction, all six categories are considered to be of equal relevance, one cannot deny a certain image-hierarchy between the different categories. In contrast to the prestigious and financially lucrative Category II (national park), the Category V (protected landscape) receives only little attention. The reasons for this are multifold.

- The outstanding image of national parks as the 'premium category of the protected areas'
- The stringent legal and spatial planning rules underlying the national parks (for example statutes instead of regulations, zoning of activities)
- The supra-regional competence of a governmental administrative body
- The differently weighted overriding management objectives.

National parks are considered to be 'majestic wilderness areas', where free rein is given to nature (see Sandilands 2000). In contrast to North America, only few national parks in Europe are attributed this sublimity on a regional level (for example Sarek National Park in Sweden). Europe's national parks frequently offer a differentiated sublimity, mostly with regard to a regionally significant cultural landscape. However it is conceded that an accurate classification of the different national protected areas has its limits and possibly has to happen even arbitrarily (see Europarc and IUCN 1999).

The 'lack of clarity' with regard to the national and international categories of protected areas can be especially well exemplified with the Category V. According to the IUCN this category describes a protected area, 'the management of which is mainly oriented towards the protection of a landscape or a marine area which also serves recreation' (Europarc and IUCN 1999, 30). It is an area 'where the interaction of man and nature has formed a landscape of a particular character over the course of time, with outstanding aesthetical, ecological and/or cultural values and oftentimes exceptional biological diversity. The undisturbed continuation of this traditional interaction is vital for the protection, conservation and enhancement of the area' (ibid.).

The disparity of large protected areas that fall into Category V is substantial. The classifications according to national law include Parco Naturale Regionale (Italy), Parc Naturel Régional (France), Parque Natural (Spain and Portugal), Naturpark (Austria and Germany), Regionaler Naturpark and Naturerlebnispark (Switzerland), Area of Outstanding Natural Beauty and National Park (Great Britain). Lastly, cross-nationally the category of the biosphere reserve is also part of Category V. Thereby the multitude of terms does not just reflect different linguistic views, but also cultural, legal and, most of all, conceptual views. For comparison purposes in literature see Henderson 1992, Schmidt 1995, Weixlbaumer 1998 and 2001, Mose and Weixlbaumer 2002, Hammer 2003.

Interpretation and application of the management categories for protected areas in Europe differ from those of other continents, for example North America. From a landscape-ecological point of view ecosystems have different types of 'nativeness', which is reflected according to Plachter (1991, 242) in the degree of hemeroby – the intensity of cultural effects on vegetation. Thus the following stages of increasing anthropogenic influence are distinguished (generalized): ahemerob (not affected – IUCN-Category I), mesohemerob (affected – IUCN-Category II), polyhemerob (strongly-affected – IUCN-Category V) and metahemerob (areas without plants). Ahemerob vegetation zones and the corresponding ecosystems are much more common in North America than in Europe. This is reflected in the continentally diverse distribution of the strict nature reserves and wilderness areas of Category I on the one hand and the protected landscapes of Category V on the other hand.

Thus, the large national parks in the United States and Canada, as opposed to European nature parks, have in many cases a dominant ahemerob portion of the surface areas. They fall only in a limited way into Category V in light of the above definition (for example in terms of a few zones), whereas all European nature parks of this category fall entirely under the definition. The different degree of hemeroby is an important reason why, in contrast to North America, the European areas of the IUCN-Category V represent the lion's share of the protected areas.

The large protected areas that fall into Category V of the IUCN, are not just of interest because of their large surface ratio, but also due to their underlying conceptual views. It is widely recognized that biosphere reserves, nature parks, regional parks and so on are considered to be the essential category of the dynamic-innovation paradigm. This paradigm is characterized by a moderate anthropocentrism, where man plays an essential integrative role. Because Category V has only a little overlap with the national park-concept, this category is, for a sustainable landscape development, an important link between polyhemerob and ahemerob landscape protection on an international level (see Weixlbaumer 2005b).

Against this background the following paragraphs dwell initially on the historical development of area protection in Europe up to the present, as well as on the resulting effects on the formulation of the main paradigm strands of the international area protection policy.

## Historical Development of Area Protection in Europe

In Europe area protection as part of nature protection – and the latter again as an element of environmental protection – is, with its term as well as with its objectives, clearly less than a century old. Nevertheless, area protection has deep historical roots. Several pieces of evidence exist to show that provisions have long been in place to protect nature and to designate protected areas (see Jäger 1994).

An illustrative example is the 8th and 9th century conservation of extensive woodlands through the creation of protected forests. These efforts were initiated by the sovereign claims of rulers in an effort to conserve forest with its resources for hunting and timber in an adequately large surface area. Starting from these early approaches, there followed an ongoing process of improvement toward today's forestry legislation. Interestingly enough, many former protected forests are conserved as 'protected areas' to this day. In Germany they frequently form the nucleus of numerous nature parks, national parks and biosphere reserves in the low mountain ranges (for example Pfälzer Wald, Bayerischer Wald, Kellerwald).

These early forms of area protecion need to be distinguished from the modern development of nature protection and hence, the resulting designation of (large) protected areas in Europe.

The origin of modern nature protection dates back to the 18th and 19th centuries. In the course of the industrial revolution and its after-effects, nature and landscape experienced a new – positive – appreciation, which was diffused gradually in society. After some time, this diffusion began to have an impact on politics, which can be seen in the incremental creation of a legal basis for nature protection from the end of the nineteenth century onward.

The primary motives for the development of the idea of nature protection and the notion of protected areas are characterized by a significant heterogeneity, which is oftentimes overlooked (see Blunden and Curry 1990, 6pp.; Erz 1994; Mose and Weixlbaumer 2003; Vogt and Job 2003; Frohn and Schmoll 2006). The following origins in intellectual history are especially relevant in this context:

- An essential incitement for the notion of protection was presented by intellectuals and artists (painters, writers) who stressed the aesthetic motives dedicated to the conservation of 'the beauty of the landscape' (example: William Wordsworth's 'Guide to the Lakes', 1810). These were articulated primarily by the middle class who saw the rural cultural landscape (in England described as the praised 'countryside') as an idealized antithesis to the urban world and industrialization.

- Alongside these aesthetic arguments for nature protection, ethical motives developed. These were driven by a belief that landscape, animals and plants, as God's creation, should be protected for their own sake. This movement that was echoed in equal measure in the middle and working class, established several nature protection organizations that have maintained their relevance to this day (for example Royal Society for the Protection of Birds, 1889; Deutscher Bund für Vogelschutz, 1899).

- From an academic environment arose a movement for scientifically founded protective measures that came along with the progression of ecology as a discrete scientific discipline (1866 definition of ecology by Rudolf Haeckel).

- Frequently overlooked are motives for protection that were verbalized previously in a proletarian setting, with the interest in nature focused primarily on the creation of nature-oriented recreational areas for the workforce. This movement gained momentum in several European countries during the 1920/30s, growing to a significant size. Important national and international leisure organizations have part of their roots in this movement (for example Naturfreunde, Alpine Clubs).

The outlined progression of the notion of nature protection in Europe corresponds closely with the one in North America, particularly in the United States. The designation of the first national park in the area of the Yellowstone River in 1872 is frequently cited as 'the trigger' for respective demands in Europe, even though the creation of the large protected areas in America took place in a completely different historic and societal context. Primarily these designations took place because of the romanticizing of archaic national landscapes in an absence of historical landmarks in the young American nation. Interestingly enough, this nation developed the idea of a national park at what was once called the 'frontier', a place of great challenges for American settlers bordering on great wilderness. The first specific implementation of these ideas happened in the Yosemite Valley, discovered in 1851 in the Californian Sierra Nevada, which was already under protection in 1863 by means of its own law as a park 'for public use, resort, and recreation' (then in 1879 formally declared a national park) (see among others Trommer 1993).

The model for national 'wilderness parks' that was developed in Yosemite under the significant influence of the conservationist John Muir, quickly became popular in the natural landscapes of North America (Yellowstone, Grand Canyon, Banff and so on). In comparison, this model was less applicable in densely populated Europe, even though shortly after several conservation movements adopted the resolve to create extensive protected areas, especially national parks following the American example. Thus, already in 1898 in the Prussian House of Representatives the demand to create reservations for nature in the form of 'state parks' was postulated.

Hence, it is easily understood that the first European national park was established in the sparsely populated north of Sweden in 1909, and that soon after further national parks in the largely uninhabited high mountains of the Swiss Engadin valley (1914), the Picos de Europa mountains in North Spain (1918) and the Italian Gran Paradiso area (1922) followed. Even in Germany a comparatively densely populated country, the first extensive protected area was set up. In this case the initiative of a private organization, the 'Verein Naturschutzpark' (1909), was decisive. Their effort resulted in the establishment of the nature protection park 'Lüneburger Heide' in 1921 with a size of 20.000 ha (today 234.400 ha); however, the objective of this park was the protection of a historic cultural landscape and not of a wilderness area (see Cordes 1997, Blab 2006).

An adequate legal foundation for nature protection, especially area protection and the designation of extensive protected areas, was implemented comparatively late. Not until 1935 did the German 'Reichsnaturschutzgesetz' (the nature protection law of the empire) come into force, and in 1949 in Great Britain the 'National Parks and Access to the Countryside Act' respectively. The latter represents the postwar development of nature protection in Europe, which is characterized by a broad legal foundation (see Figure 1.2). Along with this came a more or less gradual designation of protected areas that focused primarily on small-sized protected areas, but also on types of protected areas that had very few restrictions. In Germany this trend can be seen in the designation of several nature parks since the late 1950s, and respectively, in Great Britain the establishment of numerous national parks. It was only in the course of the European Year of Nature Protection (1970) and the Conference on the Human Environment in Stockholm (1972) that the demands of area protection experienced a significant political revaluation, which, amongst other things, became manifest in the creation of the first national parks in Germany (1970, Bayerischer Wald) and in Austria (1981/1984/1991, Hohe Tauern). Since then a downright boom to designate extensive protected areas has taken place (see Weixlbaumer 1998, 49pp.).

The recent history of area protection can be primarily characterized by the progressing internationalization of area protection-policy. This is the case as to the designation of numerous new biosphere reserves and parks respectively, especially since the Sevilla-Strategy was passed in 1995 (see ÖAW 2005) which gave UNESCO's Man and the Biosphere Program (MAB) additional significance. Thereby some large protected areas already in existence took on the additional title and designation as a biosphere reserve. In Germany, the unification with the former GDR played the role of a catalyst as numerous biosphere reserves in East Germany entered as a 'socialistic legacy' into the German nature protection law. In contrast, in other European countries the biosphere reserves date back much further. Finally there is also significant influence on the part of the EU today. In the centre of it all is the European protected areas-system NATURA 2000, which is based on the Birds Directive (1979) and the complementing Habitats Directive (1992). Hereby, for the first time on the level of the EU, there exists a comprehensive legal body for habitat and species protection (see BLAB 2006, 10). On the one hand, because of this, the

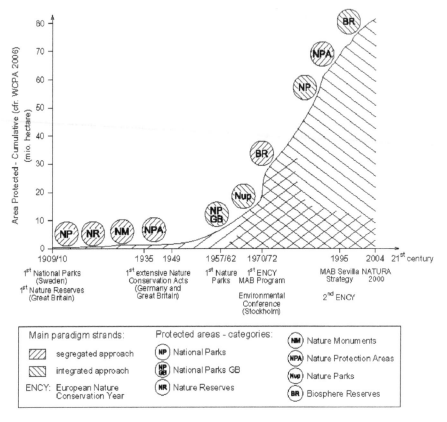

**Figure 1.2    Advancement of protected area policy in Europe**
*Source: Drafted by authors*

European realm of protected areas has become more integrated, while on the other hand it has grown to be even more complex than previously.

## Paradigm Shift in Protected Area Policy?

Depending on the latest trend of the understanding of nature and the zeitgeist respectively, different basic principles in area protection policy have developed within Europe and beyond (for example North America). Taking national park policy as an example, Henderson (1992) analyzed substantial characteristic differences between the United States and Canada as well as Great Britain. Referring to the nineteenth century, he basically distinguishes between the preservationist-movement – 'protection without use' – in the United States and the conservationist-movement – 'protection through use' – in Canada and Great Britain. On the one hand wilderness was conserved, and on the other hand natural landscape was cultivated and value was added by tourism. Preservation and protection with little understanding for integration, that is without seriously thinking beyond the boundaries of the protected area, were the basic elements of the concepts of protected areas until the middle of the

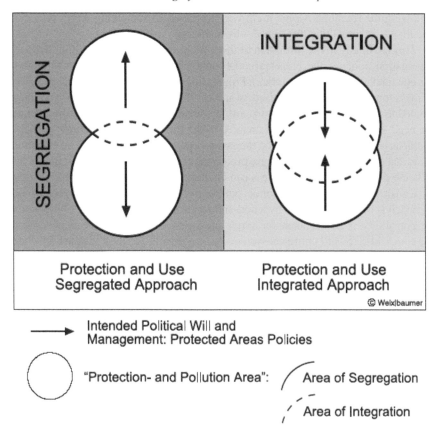

**Figure 1.3　The main paradigm strands in area protection policy**
*Source: Drafted by authors*

20th century. By the end of the last century things had changed, at least in the United States and Canada, decisively. Today the trend in both countries is headed towards ecosystem-based management (see Slocombe and Dearden 2002, 297pp.), even though this term is not always explicitly used. The ecosystem-based management approach replaces isolated nature protection with an integrative approach, which was also expressed in the Durban Accord (2003) (see Figure 1.3): 'A fundamental premise of ecosystem management is to turn protected area management from "boundary thinking" ... to an understanding of the spheres of influence that affect parks beyond the administrative boundary' (ibid., 302). In Europe, due to the cultural diversity and the developments in tourism, there are various and – just like in North America – parallel existing styles. In Austria and in Western Central Europe, from a general point of view the preservationists prevailed at the turn of the 19th/20th century. This lead to several species protection laws and the establishment of a few very strictly protected areas (for example the first protected areas Hohe Tauern – Hinteres Stubachtal 1913, Schweizerischer Nationalpark 1914). After, in the Alpine region, the notion of nature as 'Montes horribilis' was overcome, nature following the romantic Alpine myth was perceived as something worth protecting and finally as something threatened – by mankind.

This led finally to the formation of the static-preservation approach. Initially, the segregation policy – the preservationist-movement – won recognition throughout Europe (and also in North America). Progressing industrialization and the beginning of mass tourism increasingly raised awareness that natural landscapes are no longer abundant, but quite to the contrary are abundantly threatened. From this resulted the need to designate 'protection areas' – in the form of nature reserves – which is characteristic for the objective of the static-preservation approach. The remaining areas that do not enjoy a legal nature protection regulation or designation run the risk of being degraded. Thus, such areas can be described as 'pollution areas'.

Consistent with this dichotomic perspective on space, the static-preservation approach is being determined by a non-anthropocentric understanding of protection. Responsibility is being taken for single creatures (biocentrism) or, depending on the ethics, also toward entire ecosystems (ecocentrism). Because they are attributed their own intrinsic value, they deserve to be treated as morally worthy to-be-considered objects (see Stenmark 2004). Nature ought to be protected; humans take in the outsider-perspective towards their protected objects. In the course of the modern advancement of this approach the environment has become the centre of reference for a feeling of personal distance. Several phases of radical to moderate non-anthropocentrism alternate. In nature protection policy, which in many cases embodies a preservation policy by means of species and area protection, this paradigm is characterized in such a way that people's understanding of nature as well as their decision making mechanisms are on the sidelines (for example the designation of protected areas excludes bystanders). In order to additionally protect single species the focus was often on preserving entire landscapes from anthropogenic influence. Protected areas were established as wilderness or nature reserves. Examples of this are, as aforementioned, the establishment of Yellowstone National Park (1872) in the United States as well as, in Europe, the Sarek National Park in Sweden or the Swiss National Park at the outset of the 20th century.

Based on this background, one has to conclude that there are essentially two different concepts of area protection that can clearly be separated. On the one hand is the paradigm of the static-preservation approach (segregation approach) and on the other hand the paradigm of the dynamic-innovation approach (integration protection) (see Weixlbaumer 2005b).

The following attributes characterize the static-preservation approach:

- Based on the dichotomy of 'protection- and pollution area', nature protection and business areas are spatially separated that is bell jar and reservation policy respectively: protected areas have only little contact with the outside world. It is a matter of sectoral protection that only targets certain species and is, subsequently, area-oriented. The two main tools of classical nature protection are species and area protection. The mechanistic worldview is the ideological basis of this approach.
- The basic principle of species and area protection is often pursued with a rather rudimentary management structure, that oftentimes is situated only on a supra-

regional level (for example NGO or national government office). Furthermore, the management frequently has no choice but to view this task as one of many others. Norms exist – management however plays no essential role with the exception of national parks of the Category Ia and Ib. Nature protection happens through idealizing – primarily to preserve the subjects of protection.

- The shaping of norms and the designation of protected areas happens top-down. It is a sort of 'sovereign' nature protection, mostly prohibitive in nature.
- The acceptance of all interested parties (for example abutters to the protected area) is not scrutinized in this normatively driven approach.

In contrast, the paradigm of the dynamic-innovation approach (integration approach) is characterized by the following basic principles and attributes:

- Nature protection is a spatially and temporally basic principle that attempts to overcome the 'protection and pollution area' dichotomy.
  The basic principle of sustainable development is expressed by turning from pure area protection towards procedural protection. Beyond that the principle is expressed by the aimed model effect for areas and procedures outside the protected areas. The transactionistic world view is the ideological basis of this approach.
- Integrative protection and landscape development measures are usually achieved by an adequate management structure (on location and oftentimes additionally by a supra-regional alliance or a governmental office). Nature protection becomes more and more professional.
- Nature protection using a policy-mix is considered to be a societal task (top-down and bottom-up approaches are intertwined). Therefore measures are less normative, and are built on a high degree of voluntariness.
- The acceptance of all relevant people is essential. In general it is the result of a cooperative effort of all parties involved.

The background to this dynamic-innovation approach, as opposed to the static-preservation approach, is the idea of cooperation (in the sense of Meyer-Abich 1990). A moderate anthropocentrism, declining any radical forms, has priority over a non-anthropocentric view. Nature can only be protected and advanced by man in a sustainable way if mankind considers itself to be a part of nature. In this way, mankind takes on the inside perspective to its protected areas. Therefore the criticism of 'science-obsessed nature protection' does not apply to this approach (see Plachter 1991). In fact, research and management have to be oriented in an inter- and trans-disciplinary way in order to give enough consideration to integrative basic principles and attributes. A stronger recognition of the human science component is explicitly required (cf. Erdmann 2000). The outcome of this is the notion of area protection, which in Europe is above all oriented towards the sustainable development of strongly-affected landscapes, with an explicit innovation element. The main tool of this approach is in many cases the large family of 'nature parks' (and the IUCN-Category V respectively). Even if in reality this generalized polarization is not all that black and white, one can more or less describe the traditional area protection policy as being a 'protection and pollution area ideology'. Due to an increase in knowledge in the nature protection realm, the dynamic-innovation approach has been established particularly in Europe in recent years. The challenge of this approach is to make the integration of different interests (in use) possible. This approach attempts to satisfy the interests of protection and use on the 'experimental ground' or 'laboratory' nature park in a socially, economically and environmentally sound way. In other words it is the goal to simultaneously form the regulatory, habitat, support, development and information functions of protected areas, while using large protected areas beyond their boundaries as tools of sustainable regional development.

In the practical everyday world of area protection, the strategic measures and main paradigm strands presented in this paper interact in a complementary way. Depending on the requirements of nature protection and the regional situation, both approaches can be legitimately used. In this respect this is, strictly speaking, not a shift but an extension of paradigm.

Based on its basic principles, attributes and 'background-philosophies', Europe's booming nature, regional and biosphere parks, as well as numerous national parks (that are not part of the IUCN-Category II) of the past decades have to be attributed to the dynamic-innovation approach. If one excludes national parks, there are roughly 600 managed protected areas that fall into the IUCN-Category V in Europe alone. The total surface area of all IUCN-Category V areas comprises 345,821 $km^2$ (see Figure 1.1).

## Nature Parks as Model Landscapes

Against the background of the discussion of a paradigmatic advancement of nature protection, today large protected areas are often considered to be 'model landscapes', and they are not uncommonly seen as tools for regional development. This is especially the case for managed protected areas where there is considered to be an immense stimulating potential, particularly for the advancement of peripheral rural areas (see among others Mose/Weixlbaumer, Hammer 2003). The terminology, but also the substantive positions in relation to the large protected areas and regional development are however very different and convey a rather fuzzy picture. As Hammer explained in chapter 2, there are significant differences as to whether protected areas ought to be considered or used as 'pulse generators', 'engines' or 'tools' of regional development.

For the recent paradigmatic advancement in area protection, nature parks of the IUCN-Category V represent an especially interesting example. Because no other protected area category represents so well the advancement of the European area protection policy and with it the resulting paradigm shift, but at the same time also the underlying developmental path of the mankind-nature(protection)-relationship in Europe. The nature parks have a paradigmatically double connotation. Thus, many nature parks combine the objective of preserving protection – the legal starting point of every nature park in Austria and Germany is a nature or landscape protection area – as well as the notion of the dynamic-innovation 'advancement-protection'.

In the following the advancement of the nature park-functions will be exemplified on the basis of the experiences of Austria (see also chapter 5).

At the time when nature parks were established in Austria, they were nothing but local recreational landscapes for Vienna's urban agglomeration. The recreational function of 'nature' – as a large garden-landscape – was very modestly combined with nature and landscape protection. This can be exemplified with the first nature parks in Lower Austria during the 1960s (see ÖGNU and Wolkinger 1996). Within the scope of its local recreational function the nature parks enjoyed increasing offers of nature and environment education for the broad population. Thus, nature parks turned gradually into 'educational landscapes' by the setup of various nature trails and information panels. At the outset the pedagogical implementation of nature and environment education was modest. At first, there was also a lack of appropriately designated landscapes – these few small nature parks represented the Austrian landscape types inadequately. In addition there was also a lack of political willingness, which is apparent in the marginal funding and support. Moreover the local residents were only sparsely integrated in nature park policy. Overall there was a lack of awareness for integrated nature and landscape protection.

The activities were intensified and coordinated only as a result of the institutionalization of nature parks through the founding of the association Austria's Nature Parks (1995) as well as the international appreciation of this protected area-category. The coexistence of the functions of nature protection, recreation and education, which in the meantime had been elevated to a credo, was soon after extended. From then on, the landscapes of nature parks were also obliged to contribute to the regional development of rural areas. Corresponding to the paradigmatic advancement, the protection, recreation, education and regional development functions were subsequently seen to cooperate. Today nature parks are considered to be an element for integrated regional development in rural areas. Initially in this respect the federal state Styria proved to be quite innovative within Austria. It not only established, in a relatively systematic way, extensive nature parks, but also provided a solid infrastructure for these areas. This includes minimum budget, a director (independent from the municipality), a nature park academy, public relations and so on Upper Austria in turn promotes nature park development from the region which is bottom-up (for example nature park Rechberg/Mühlviertel). Hence it only has a few, but well functioning model nature parks. In the future, nature parks in Upper Austria will seek to include all significant types of cultural landscapes of the country. The regional management of Burgenland pursues a similar policy with its own concept. Furthermore, Tyrol is an innovator with its current concept of managed protected areas (example: High Mountains Nature Park Zillertal). With it, one can recognize an increasingly systematic approach in Austria's nature park policy.

In the meantime, nature parks throughout Europe have become research objects with respect to their potential for the advancement of marginalized rural areas. At least for the proponents of nature parks, the image of a sustainable advancement of natural and cultural landscape was one of the key interests for establishment. An image that ideally should radiate from the nature parks on to other landscapes. A set of basic conditions is required however, to satisfy the claim to use nature parks as real 'pulse generators' of regional development.

In order to add a value to the normally abundant qualities of the nature park regions – as model landscapes of a country – it is essential to have the broadest possible involvement of regional and local players. This includes the local residents in particular. An adequate policy-mix with respect to an endogenous regional development – albeit with exogenous support – offers the prerequisites to implement communicative and participatory forms for the planning of protected areas and thus ensures a high level of acceptance among bystanders. The implementation of the EU-initiative LEADER exemplifies in particular the possibilities of such participatory bottom-up development strategies in different nature park regions.

Furthermore, to implement a regional development that is suitable (that is integrated) for a nature park, there needs to be a relevant political will. Without this will, every kind of planning for the protected area is likely to fail. Primarily political will is expressed by the enforcement of (nature protection) laws. But it also means using existing networks of different stakeholders and involving public institutions, business people and players of civil society simultaneously in the spirit of modern governance (see Fürst et al 2005).

Overall, it is about arranging and implementing an entire set of fundamentals, which include a legal, planning and financing foundation. Build on this, one has to ensure the efficiency of the park by means of an adequate surface area, if applicable, zoning, as well as a sufficient staff. As a rule of thumb, for the park staff one generally counts one person per 2,000 ha depending on the Alpine landscape (see Scherl 1989). Thus, a nature park can become an efficient enterprise for a region. In addition, (depending on the season) a certain number of volunteers and part time workers is required, the exact number variable based on the needs of every park.

In conclusion, a nature park shall not only be seen as a single, isolated protected area, but also as an element of a larger network of regions. Its unique potential needs to be regarded, cared for and as the case may be, a value needs to be added. The positive impact beyond park limits – by means of innovations in the areas of organic farming, environmentally and socially compatible tourism or the use of renewable energy – need to be treated just like all occurrences in the park itself, as a building block of a network, and thus require targeted support. Nature parks can thereby surpass the conventional evaluation procedures and become 'learning regions'. At the same time they stand for 'teaching regions' by using lived, innovative sustainability-strategies. In sum, nature parks have considerable potential to be the pulse generators, the desired landscapes or even the model landscapes for sustainable development.

## Outlook

In the last two decades large protected areas have increased their importance within Europe. Currently, it is experiencing yet again an accentuation. This is not just true because of the substantial surface ratio in various European countries, but is also true with regard to the diverse functions of large protected areas. In connection with this is the addressed paradigm shift (paradigm complement). Accordingly the advancement function of protected areas has received an increasingly significant importance as opposed to the protection function. An immediate expression of this fact is the downright boom of protected areas of the IUCN-Category V, which are all oriented towards the integration of protection and use functions. In Europe this is, particularly on a regional level, the very heterogeneous nature and regional park setting. On an international level one has to mention the prestigious biosphere concept. Examining its functions of (according to UNESCO) development, conservation, as well as logistical support, one can see that it represents the implementation of the sustainability notion into practice (see ÖAW 2005).

For the future advancement of protected area policy it will be important to primarily pursue the path of stronger integration of protection and use in the years to come. In addition, large protected areas ought to be used consistently (also) as tools for a sustainable regional development. As the previous practical experience shows, the approach of a dynamic-innovation area protection seems to be an increasingly suitable conceptual framework which can highlight practicable models for mankind to treat the landscape that we live in and that we use, be it as an individual or as a society. In short this framework should enable us to try out sustainable ways to live and do business.

The challenges for the future area protection policy in Europe remain enormous, despite the achieved successes. The consensus among many experts is that the objectives and the reality of modern area protection still diverge significantly. Accordingly, there remain numerous tasks both to the extent and the quality of the protection, in order to constitute and establish ideal protected area-systems under the given basic conditions – particularly in the densely populated countries of Central Europe. In doing so, the intelligent differentiation and combination of different concepts, protected area-categories and levels of activity is going to be of decisive importance. This precisely means to intelligently connect the static-preservation elements with the dynamic-innovation elements of area protection. The different types of protected areas such as national park, nature park or biosphere reserve, to mention only three central categories, can be combined to protected area-systems that are spatially differentiated and regionally adjusted. Likewise the opportunities arise to differentiate within the different categories of protected areas, which is already often practiced with the conventional zoning-concepts. Depending on the size and the function attributed, there is a gradual hierarchy of the protected areas implied – which spans from a regional to national to international level of importance (see Ssymank 1997, 35–6).

Against this background it becomes clear that in the upcoming years large protected areas will be increasingly confronted with the claim to serve as tools of regional development. This claim has an even stronger effect, inasmuch as it is combined in a paradigmatic way both with the large number of IUCN-Category V areas in Europe and the biosphere reserves of the second generation (after Sevilla 1995). This high expectation from protected areas to combine nature protection and regional development in a sustainable way is at the same time legitimate and dangerous. As Hammer (2003, 30) appropriately noticed, there is a lurking danger that if the expectations are persistently kept inflated, those protected areas that will not meet these expectations will end up in a losing position. Consequently, protected areas might end up collectively losing acceptance. Thus, neither the objectives of regional development nor of nature protection will be served. In this respect there remains the hope that protected areas are more than just life-savers in the regional economic and/or ecological struggle for survival.

**References**

Blab, J. (2006), 'Schutzgebiete in Deutschland – Entwicklung mit historischer Perspektive', in *Natur und Landschaft* 81: 1, 8–11.

Blunden, J. and Curry, N. (eds) (1990). *A People's Charter* (London).

Büchter, C. and Leiner, C., (2000), *Schutzkategorien im Naturschutzrecht. Systematische und kritische Übersicht* (Kassel).

Chape S. et al. (2003). *2003 United Nations List of Protected Areas.* IUCN Gland, Switzerland and Cambridge, UK and UNEP-WCMC (Cambridge, UK).

Cordes, H. (Ed.) (1997). *Naturschutzgebiet Lüneburger Heide. Geschichte – Ökologie – Naturschutz* (Bremen).

Erdmann, K.-H. (2000). 'Naturschutz – quo vadis? Anregungen zu einer Neuausrichtung'. In: *PGM*, 143, 80–85.

Erz, W. (1994). *Geschützte Natur nach dem Bundesnaturschutzgesetz.* In: Praxis Geographie 12, 7–12.

EUROPARC; IUCN (1999). *Richtlinien für Management-Kategorien von Schutzgebieten. Interpretation und Anwendung der Management-Kategorien für Schutzgebiete in Europa* (Grafenau).

Frohn, H.-W. and Schmoll, F. (2006): 'Amtlicher Naturschutz – Von der Errichtung der „Staatlichen Stelle für Naturdenkmalpflege" bis zur „ökologischen Wende" in den 1970er Jahren'. In: *Natur und Landschaft* 81, H. 1, 2–7.

Fürst, D. et al. (2005): 'Regional Governance bei Gemeinschaftsgütern des Ressourcenschutzes: das Beispiel Biosphärenreservate'. In: *Raumforschung und Raumordnung* 5, 330–339.

Hammer, Th. (ed.) (2003). *Großschutzgebiete – Instrumente nachhaltiger Entwicklung* (München).

Henderson, N. (1992). 'Wilderness and the Nature Conservation Ideal: Britain, Canada and the United States Contrasted'. In: *AMBIO*, 21, 6, 394–399.

Jäger, H. (1994). *Einführung in die Umweltgeschichte* (Darmstadt).

Lillo, A.L. (1995). 'Europäische Regional- und Naturparke'. In: *Europäisches Bulletin. Natur- und Nationalparke* 125, 28–31.

McNamee, K. (2002). 'Protected Areas in Canada: The Endangered Spaces Campaign'. In: Dearden, P. and Rollins, R. (eds). *Parks and Protected Areas in Canada. Planning and Management* (Oxford, N.Y.), 51–68.

Meyer-Abich, K.M. (1990). *Aufstand für die Natur. Von der Umwelt zur Mitwelt* (München).

Mose, I. and Weixlbaumer, N. (eds) (2002). *Naturschutz: Großschutzgebiete und Regionalentwicklung* (= Naturschutz und Freizeitgesellschaft, Bd. 5) (Sankt Augustin).

Mose, I. and Weixlbaumer, N. (2003): 'Großschutzgebiete als Motoren einer nachhaltigen Regionalentwicklung? – Erfahrungen mit ausgewählten Schutzgebieten in Europa'. In: Hammer, Th. (ed.) 2003. *Großschutzgebiete – Instrumente nachhaltiger Entwicklung* (München), 35–95.

ÖAW (Österreichische Akademie der Wissenschaften) (ed.) (2005). *Leben in Vielfalt. UNESCO-Biosphärenreservate als Modellregionen für ein Miteinander von Mensch und Natur* (Wien).

ÖGNU and Wolkinger, F. (eds) (1996). *Natur- und Nationalparks in Österreich* (Graz).

ÖROK (Österreichische Raumordnungskonferenz)(ed.)(1997).*Naturschutzrechtliche Festlegungen in Österreich*, Schriftenreihe Nr. 135 (Wien).

Plachter, H. (1991). *Naturschutz* (Stuttgart).

Revermann, Ch. and Petermann, Th. (2003). *Tourismus in Großschutzgebieten. Impulse für eine nachhaltige Regionalentwicklung* (= Studien des Büros für Technikfolgen-Abschätzung beim Deutschen Bundestag, 13) (Berlin).

Sandilands, C. (2000). 'Canada's National Parks: Profits, Preservation and Paradox'. In: *Canadian Parks and Recreation*, 57, 6, 16–17.

Scherl, F., Tome, A., Broili, L., Perco, F. and Perco, F. (1989). *Parco Naturale delle Prealpi Carniche. Piano di conservazione e sviluppo.* D.6: Norme per l'esecuzione del piano (Trieste).

Schmidt, G. (1995). 'Naturschutzplanung in Spanien. Verbindung zwischen Naturschutz und umweltverträglicher Entwicklung – ein Modell für die ländlichen Räume Europas?' In: *Naturschutz und Landschaftsplanung* 27, 69–75.

Slocombe, S. and Dearden, Ph. (2002). 'Protected Areas and Ecosystem-Based Management'. In: Dearden/Rollins (eds). *Parks and Protected Areas in Canada. Planning and Management* (Oxford, N.Y.) 295–320.

Ssymank, A. (1997). 'Schutzgebiete für die Natur: Aufgaben, Ziele, Funktionen und Realität'. In: Erdmann, K.-H. and Spandau, L. (eds). *Naturschutz in Deutschland. Strategien, Lösungen, Perspektiven* (Stuttgart), 11–36.

Stenmark, M. (2004). 'Überblick über einige normative Ethik-Prinzipien von Biozentrismus und Ökozentrismus'. In: *Natur und Kultur*, H. 2, 88–113.

Trommer, G. (1993). 'Nationalpark und Biosphärenreservat. Zum Landschaftsbezug in der Umweltbildung'. In: *Geographie heute* 115, 4–8.

WCPA (World Commission on Protected Areas) (2006). *Protected areas growth for region: Europe*, http://sea.unep-wcmc.org/wdpa/statistics.

Weixlbaumer, N. (1998). *Gebietsschutz in Europa: Konzeption – Perzeption – Akzeptanz* (=Beiträge zur Bevölkerungs- und Sozialgeographie, 8) (Wien).

Weixlbaumer N. (2001). 'Gebietsschutzpolitik in Italien und Frankreich – Welcher Umgang mit welcher Natur?' In: *Berliner Geographische Arbeiten*, 91, 97–105.

Weixlbaumer N. (2003). 'Alpenschutz'. In: *Bedrohte Alpen, Segmente – Wirtschafts- und sozialgeographische Themenhefte*, 33–41.

Weixlbaumer, N. (2005a). 'Landschaftswünsche werden wahr. Wie Naturbilder die Landschaft prägen'. In: *umwelt & bildung*, H. 1, 11–13.

Weixlbaumer, N. (2005b). '"Naturparke" – Sensible Instrumente nachhaltiger Landschaftsentwicklung. Eine Gegenüberstellung der Gebietsschutzpolitik Österreichs und Kanadas'. In: *Mitteilungen der Österreichischen Geographischen Gesellschaft* 147, 67–100.

Chapter 2

# Protected Areas and Regional Development: Conflicts and Opportunities

Thomas Hammer

The paramount objective of a protected area is, according to the internationally accepted definition, conservation. A protected area is an area of land and/or sea especially dedicated to the protection and maintenance of biological diversity, and of natural and associated cultural resources, and managed through legal or other effective means (see IUCN 1994). Nevertheless, protected areas are increasingly being viewed in the context of regional development expressly for the sake of achieving conservation objectives. It is increasingly broadly accepted that co-ordinating conservation and the utilization of nature is advantageous for both conservation and regional development. The supposed antithesis between use and conservation, which is played up time and again in the media in relation to tourist infrastructure projects, seems to be breaking down. The idea of combining conservation and regional development is gaining impetus, particularly in the discussion of sustainable development.

In this chapter we will show how this re-evaluation of the connexion between conservation and regional development came about and what the theoretical approaches to this integrative view are. We will also clarify what opportunities exist for integrating the goals of conservation and regional development and where the limits to this lie and we will look at how regional policies support this integration. Finally, the challenges to protected areas and to research on protected areas will be outlined in the context of the discussion on regional development.

## Conservation and Regional Development – Partners or Field for Conflicts?

As a matter of principle, conservation, and in particular the implementation of protected areas, is always associated with conflict. Conservation only becomes necessary when one or more actors wish to protect parts of nature from human use. A particular activity is either desired or prohibited, thus restricting the freedom of action of one or more other actors. In protected areas entire tracts of land are withdrawn from human use or regulations limit the activity of the actors and require a particular activity. Conservation always restricts human freedoms, and this inevitably encourages actors to oppose the regulations. In this sense conflicts – and likewise the regulation of conflicts – must be considered a constitutive component of conservation.

Conflicts regarding protected areas usually focus on differing interests in the use of natural resources. Whereas some actors want to preserve, care for, or enhance a particular natural resource, others want to use or exploit it. An alternative, which classical regional planning has employed for a long time, is to set spatial priorities for use. Zones are delimited in which living, work, industry, traffic, forest use, conservation, etc., takes priority.

But the classical zoning and spatial segregation of human activities employed by regional planners is not sufficient to regulate conservation, care, preservation and development on a large scale. Practically no large scale protected areas exist that are used by humans only to a slight degree or only gently. Neither ignoring the old types of use nor resettling the population are reasonable strategies for achieving the aims of a protected area. On the contrary, closing off large protected areas can prove counterproductive if the regulations for use are not adhered to and the result is arbitrary use. Moreover, the zoning of use done by regional planners is usually on a small scale and often leads to small zoning units and thus to a patchwork quilt of small protected areas. Many conservation objectives, however, such as the preservation of specific habitats, biodiversity or natural and cultural landscapes, cannot be achieved in this manner.

The type of regional planning in which different types of use are segregated divides up the region and imposes different priority uses, making it difficult to achieve especially large scale conservation objectives. Moreover, classical regional planning tends to be reactive rather than to try to direct development. It determines specifically the spatial development of settlements and infrastructure. Semi-natural regions remain residual quantities that can be incorporated into the developing settlements and infrastructure as needed, unless relevant conservation laws prevent this. Moreover, conservation is basically not a priority in regional planning, unless there are appropriate laws or regulations that make their implementation mandatory for landowners.

Regional planning is an important instrument for the implementation of conservation regulations. By itself, however, it is not sufficient. Large scale objectives are required. There must be a certain consensus regarding these objectives and relatively widely accepted regulations for use, before large scale regional planning measures can be devised and the conservation objectives can be implemented in such a manner that they are mandatory for landowners.

The crucial reason why conservation and regional development are increasingly being seen as an integrated whole has to do with a new understanding of development. If the bases of life are to be preserved long-term for later generations, a type of development is needed that does not destroy our resources to the previous extent. Zoning that distinguishes between protected areas and areas that can be used will not suffice in the long run, in view of the continuous spatial spread of human types of use and the increasing intensity of use. Sustainable development requires that resources be utilized in a natural manner that preserves the resources and views them as a stock of natural capital. Rather than depleting this capital, only the return on the capital should be used.

To summarize, particularly since the 1980s and 1990s a threefold paradigm shift has occurred:

- In the general discussion of development there is a call, expressed in the slogan 'sustainable development' for a type of human development that does not destroy natural resources, but is instead based on the return on the natural capital. Such demands derive from the assumption that if man is to survive on earth in the long run, he must live in greater harmony with nature.
- In the discussion of conservation there is a call for comprehensive, large scale conservation and not only selective, small scale or segregating conservation, because with this alone it is impossible to achieve many of the objectives of conservation. Conservation should be integrated into such fields as agriculture and forestry, hydraulic engineering, tourism and the development of settlements and infrastructure.
- In the discussion of regional policy and of the future of rural regions increasing importance is being attached to the promotion of endogenous potentials and local initiatives, that is the promotion of local resources of all types (including natural, social, cultural, economic and technological resources). One reason for this is that many of the anticipated effects of policies that distributed funds indiscriminately failed to materialize. On the other hand, the public authorities are less and less willing to make large amounts of public funds available to regions that do not undertake efforts of their own. Grassroots initiatives and personal contributions are called for, and these are then supported. In this view the natural conditions of a region become a development factor.

This threefold paradigm shift moves human beings and nature into the centre of interest in conservation and development. Humans and their types of use are seen as part of the natural and regional development. People move into the centre and are (on the other hand) recognized as part of nature. The human environment perspective (mankind as part of nature) replaces the environmental perspective (nature = nonhuman nature).

That this transformation has been in progress for some time is shown by the categories of protected areas newly defined in 1994 by the International Union for the Conservation of Nature and Natural Resources (see IUCN 1994) (see Table 2.1). A classification is, however, not a quality seal. It merely shows what the priority objectives of the management of the protected areas are. Whether the objectives are also achieved is a different question.

All six categories of protected areas have a human significance, even the *Strict Nature Reserves* (category Ia), in which scientific research and environmental monitoring are allowed. Protected areas of all categories also have a potential regional economic significance, especially because they are nature preserves and therefore stand for immaterial values (including natural beauty, attractive landscapes). This means that they contain a potential for tourism even if they do not pursue such an objective. For example, the Swiss National Park, a *Strict Nature Reserve*, contributes substantially to regional tourism, particularly in the summer season, although this is not its goal.

**Table 2.1        The IUCN categories of protected areas**

| Category and goal of the protected area management | Definition |
|---|---|
| **Ia:  Strict Nature Reserve**<br>Protected area managed mainly for science. | Area of land and/or sea possessing some outstanding or representative ecosystems, geological or physiological features and/or species, available primarily for scientific research and/or environmental monitoring. |
| **Ib:  Wilderness Area**<br>Protected area managed mainly for wilderness protection. | Large area of unmodified or slightly modified land, and/or sea, retaining its natural character and influence, without permanent or significant habitation, which is protected and managed so as to preserve its natural condition. |
| **II:  National Park**<br>Protected area managed mainly for ecosystem protection and recreation. | Natural area of land and/or sea, designated to (a) protect the ecological integrity of one or more ecosystems for present and future generations, (b) exclude exploitation or occupation inimical to the purposes of designation of the area and (c) provide a foundation for spiritual, scientific, educational, recreational and visitor opportunities, all of which must be environmentally and culturally compatible. |
| **III:  Natural Monument**<br>Protected area managed mainly for conservation of specific natural features. | Area containing one or more specific natural or natural/cultural feature which is of outstanding or unique value because of its inherent rarity, representative or aesthetic qualities or cultural significance. |
| **IV:  Habitat/Species Management Area**<br>Protected area managed mainly for conservation through management intervention. | Area of land and/or sea subject to active intervention for management purposes so as to ensure the maintenance of habitats and/or to meet the requirements of specific species. |
| **V:  Protected Landscape/Seascape**<br>Protected area managed mainly for landscape/seascape conservation and recreation. | Area of land, with coast and sea as appropriate, where the interaction of people and nature over time has produced an area of distinct character with significant aesthetic, ecological and/or cultural value, and often with high biological diversity. Safeguarding the integrity of this traditional interaction is vital to the protection, maintenance and evolution of such an area. |
| **VI:  Managed Resource Protected Area**<br>Protected area managed mainly for the sustainable use of natural ecosystems. | Area containing predominantly unmodified natural systems, managed to ensure long term protection and maintenance of biological diversity, while providing at the same time a sustainable flow of natural products and services to meet community needs. |

*Source: IUCN 1994*

The categories II (National Park), V (Protected Landscape/Seascape) and VI (Managed Resource Protected Area) additionally provide for or call for much more extensive human use:

- National Parks (protected areas of category II) are managed as places in which spiritual, scientific, educational and recreational needs can be satisfied, but always in an environmentally and culturally compatible manner.
- Protected Landscapes (protected areas of category V) are intended to preserve areas in which the interaction of people and nature has led to a special landscape and biodiversity with high aesthetic, ecological and/or cultural value. Such areas postulate past and also future human use.
- Managed Resource Protection Areas (protected areas of category VI) are managed so as to ensure long-term protection and maintenance of biodiversity and simultaneously provide a flow of natural products and services to meet the needs of the people. They aim at a balance between protection and use to ensure that the needs of the people are cared for on a long-term basis.

At the last survey and analysis of protected areas by the United Nations, these three categories accounted for approximately 65 per cent of the area of the classified protected areas worldwide. Protected areas of the categories II (National Parks) and VI (Managed Resource Protection Areas), each with around 29 per cent of the area of all classified protected areas worldwide, form the most important type from the point of view of the area they cover. There are, however, great differences from continent to continent. Whereas in North America National Parks (category II) lead the list with 37 per cent of the area covered, in Europe Protected Landscapes (category V) dominate with more than 46 per cent of the area covered (see Chape et al. 2003: 21–30). This means that protected areas that are explicitly managed for the use of people or even presuppose such use predominate from the point of view of area worldwide. Protection and use are no longer an antithesis and in some cases this has been true for a long time. On the contrary, many protected areas attempt to integrate protection and use or only came into being through such integration.

Since the *United Nations Conference on Environment and Development (UNCED)* in Rio de Janeiro in 1992 the number of protected areas has grown greatly. In 2003 the *World Database on Protected Areas* of the United Nations listed more than 102,100 protected areas with an area of more than 1000 ha (10 km$^2$). Taken together they made up about 13 per cent of the land worldwide. Whereas the area covered by protected areas increased by 53 per cent to 18.8 mill. km$^2$ between 1992 and 2003, the number of protected areas even more than doubled from 48,400 to more than 102,000 protected areas (see Chape et al 2003: 26). Of these 67 per cent were classified by the *World Conservation Monitoring Centre (WCMC)* of the *United Nations Environment Programme (UNEP)* according to the IUCN category system. The extremely rapid increase in protected areas is an indication that they are gaining in importance worldwide, and the curve is rising, even in Europe. Protected areas will continue to increase in area and provide further opportunities for integrating conservation and use.

The change from a concept of conservation that segregates to one that integrates, that is from one that excludes people from conservation to one that incorporates human needs for use, is illustrated well by the change in the UNESCO Biosphere Reserve concept. The establishment of a worldwide network of UNESCO Biosphere Reserves began in 1976 on the basis of a concept initiated in 1974. The original priority was to protect, investigate and preserve the natural ecosystems in the most important biogeographical regions of the Earth. After the Rio Conference in 1992 and the adoption of the *Convention on Biological Diversity*, which prescribes not only the preservation of biodiversity but also its sustainable use, the Biosphere Reserve concept was further developed. Since the conference in Seville in 1995, during which the so-called *Seville Strategy for Biosphere Reserves* was adopted, Biosphere Reserves (see UNESCO 1996) have three paramount objectives. Biosphere reserves should:

- Conserve natural and cultural diversity
- Be utilized as models of land management and of approaches to sustainable development and
- Be used for research, monitoring, education and training.

This means that UNESCO Biosphere Reserves have to meet high expectations. They are supposed to preserve not only the natural biodiversity, but also the culturally influenced biodiversity originating from human use of the natural resources and thus implicitly the human use systems. They should serve as a model for long-term compatible land use and sustainable development and as a model for education towards sustainable development.

Another aspect is spatial zoning to provide for graded intensity of protection and human use (see Figure 2.1). Whereas in the core zone the protection and preservation of the natural ecosystem take priority, in the buffer zone the man-made ecosystems that are worthy of preservation are to be preserved along with the forms of use leading to them. In the development zone, which encompasses the human settlements, better use is to be made of the man-made ecosystems.

Especially since the beginning of the 1990s and the discussion of sustainable development there has been a change in attitude towards the priority objectives of protected areas. This is revealed in the studies by the various organizations in charge of protected areas. For example, particularly since the beginning of the new millennium the annual conferences of the *Europarc Federation* have been dedicated to themes related to conservation and socio-economic development, for example the topics 'Young People in Protected Areas' (2000), 'Protected Areas and the Challenge of Tourism' (2001), 'Sustainable Development in the Protected Landscapes' (2002), 'Striking the Balance Between Nature Conservation and Local Economic Development' (2003) and 'Conservation and Opportunities for People' (2004). The *Snowdonia Declaration* on *Sustainable Development in the Protected Landscapes of Europe* (2002) challenges the *Europarc Federation* to view protected areas of IUCN category V (Protected Land-/Seascapes) – and these predominate in Europe – as 'living landscapes', which are supposed to reconcile conservation and the human use of resources with each other in an exemplary manner.

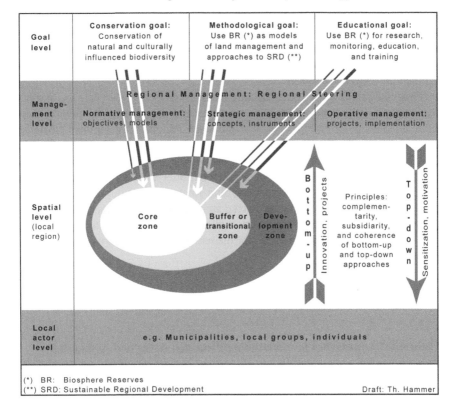

**Figure 2.1    Model of the UNESCO biosphere reserves**
*Source: Drafted by author*

**Protected Areas as Motors and/or Instruments of Regional Development?**

With the additional expectations aimed at protected areas these are increasingly being viewed as the actual motors of regional development or even as instruments of regional development (see Hammer 2003):

- Protected areas should have a positive impact on the regional economy, for example through tourism, conservation activities, the marketing of high quality regional goods and the development of regional value-added chains.
- This should also lead to positive social and cultural effects. New jobs should be created, or existing jobs should at least be preserved and made secure, thus diminishing out-migration, particularly of young persons. They should also reinforce the regional identity and improve the marketing strategy and the image of the region. In like manner they should contribute to education and open up perspectives for the future, particularly for the young inhabitants.
- Not least they are supposed to conserve the regional biodiversity, secure habitats for endangered animal and plant species and make better use of the natural, semi-natural and culturally formed landscapes.

In accordance with these high demands, the establishment of a protected area is sometimes expected to serve as a motor for regional development. They are supposed to provide important impulses for regional development, for example in tourism, in the marketing of regional products and in general in the development of innovative regional products and services.

Protected areas that were actually established as instruments of regional development have to meet even higher expectations. In this case the protected area must not only contribute to, but also influence regional development, including regional economic development. This holds, for example, for the UNESCO Biosphere Reserve Entlebuch in Switzerland, which is analysed in chapter 3.

The highest expectation is that a protected area should be an instrument of sustainable regional development according to the UN definition of sustainable development. This means that a protected area should solve regional problems and simultaneously contribute to the solution of national and international problems as formulated in the national strategies for the national level and in Agenda 21 for the global level.

Whether, to what extent and how such expectations can be fulfilled by protected areas will be shown by the case studies. For the time being the fundamental question is which theoretical approaches promise to fulfil these expectations.

## Theoretical Approaches for Combining Protected Areas and Regional Development

Just as we can distinguish between large scale protected areas that are expected to serve as a motor of regional development and ones that are employed as actual instruments of regional development, we can distinguish two degrees of intensity in the connexion between protected areas and regional development.

The connexion is of low intensity if a protected area is expected to contribute to regional development, for example by increasing the added value in the region. This kind of expectation can usually be achieved relatively rapidly. The question, however, is what constitutes an appropriate contribution. Is it sufficient for a protected area to contribute 0.5 per cent to the regional added value, or must it be 3 per cent, 5 per cent or even 10 per cent, if it is to fulfil the expectations? There is no generally valid answer to such questions.

By contrast, the connexion is of higher intensity if a protected area is to serve as an instrument for directing regional development. This completely different expectation is much more difficult to fulfil. To use a protected area as an instrument of regional development usually requires adaptations in the institutional environment. The management of the protected area must be granted the appropriate authority and funds without which this expectation cannot be met.

Whereas it is relatively easy to bring about a low intensity connexion – protected areas generally contribute to regional development in some way or other – creating a high intensity connexion requires in any case an appropriate strategic course of action. Various approaches are conceivable. In the following we will summarize four complementary approaches.

A.  Promoting networks: networks are characterized by equal partners, dialogue, negotiation, trust, reciprocity and a high degree of self-regulation. Integrating the actors into a network enables them to work together to develop ideas and design and produce new products and value-added chains within a sector of the economy or involving various sectors.

B.  Promoting creative, innovative milieus: territorial networks that continuously produce innovations can be referred to as creative, innovative milieus. Such networks are of great significance for regional development and are simultaneously difficult to control.

C.  Promoting learning regions: in this approach a 'learning region' is seen as a polycentric field of actors that is characterized by co-operative action and interactive or collective learning through face-to-face contact ('learning by interacting' or 'learning by networking'). Social networks and socio-cultural proximity represent merely preconditions but not sufficient conditions for generating knowledge, skills and creativity. Inter-organizational and interregional learning can help to maintain the process of regional development.

D.  Promoting regional cycles: from the viewpoint of sustainable development regional cycles consist:

    •  In the ecological dimension of regional flows of matter with the objective being long-term use of the natural resources
    •  In the economic dimension of regional value-added chains and
    •  In the social dimension of regional action chains, which simultaneously represent the precondition for the development of regional flows of matter and regional value-added chains.

The four approaches have in common that they characterize the region not as a political-administrative unit, but as a *social area* characterized by *relative* spatial proximity. *Social proximity* helps to mobilize the knowledge and skills of the actors and enable reciprocal inter-organizational and interregional learning and the formation of a shared identity. As a result new projects should be developed, new structures evolve and the endogenous potentials should be utilized in the dimensions of sustainable development.

A crucial question is how to actually promote networks, creative milieus, learning regions and regional cycles. Neither 'bottom-up' self-regulation nor 'top-down' regulation (external regulation) by themselves or in combination are sufficient. Instead there must be an intermediate layer and intermediate actors (between the state and the private actors). They must take over the co-ordination and integration of the self-regulation and external regulation processes, promote co-operative action, implement projects that will initiate networking and carry out long-term process management, so that the process of regional development can be set in motion and maintained over time. Likewise the opening up and renewal of the networks is an important task if an innovative process is to progress.

In the discussion about promoting the four mentioned approaches (networks, innovative milieus, learning regions, regional cycles) two concepts take centre stage, regional management and regional governance.

*Regional management* can be viewed as an approach for promoting networks, creative milieus, learning regions and regional cycles. A concrete example would be to advise the actors, organize thematic activities, initiate processes of self-organization, assist working groups, Agenda 21 processes and regional conferences and support the 'organization of self-organization'. Regional management in the sense of the above outlined approaches means guiding the collective forms of action towards results via network-like co-operative action. The intention of regional management is to contribute to the solution of regional problems. It is process, project and implementation oriented.

Whereas the discussion of regional management focuses on directing regional development along the lines of jointly developed concepts, the discussion of *regional governance* additionally deals with questions regarding mixed forms of state regulation and self-regulation and the interaction patterns between the state and society at the regional level. At the regional level there is usually no strong direct state control. It often forms only an administrative level. Since the protected areas usually spread across several communities, but are often smaller than the next higher administrative unit, additional administrative structures become necessary. If the protected area is to actively pursue regional development and not only passively manage the protected area, appropriate regional structures have to be created and the management must be granted the appropriate authority and funds without which objectives cannot be achieved. The question as to whether the protected area is embedded in adequate governance structures is particularly crucial for protected areas that aspire to a more intense combination of conservation and regional development and want to be actual instruments of regional development.

### Ideal Purposes of Regional Development in Protected Areas

Our previous argumentation leads to the following ideal expectations, especially if protected areas want to contribute substantially to sustainable regional development or are even supposed to be or want to be an instrument of regional development (see Figure 2.2).

First: protected areas consider the promotion of regional development to be their main purpose, while aiming at goals in the ecological dimension, the economic dimension and the socio-cultural dimension of sustainable development. The object is to make the best use of the regional potential in the three dimensions of sustainable development with as little conflict as possible.

Second: to achieve their objectives the management of the protected area must be granted the commensurate authority, and the level of authority must be appropriately high, if the protected area is employed as an instrument of regional development. The protected area must become an intermediate level that can co-ordinate activities on the regional level and simultaneously between the regional level, the local level and the higher levels. The management of the protected area must, for instance, be able to create incentives for local actors and to solicit support from a higher level.

Third: the protected area must be able to make substantial contributions on the normative, strategic and operative levels. The management of the protected area

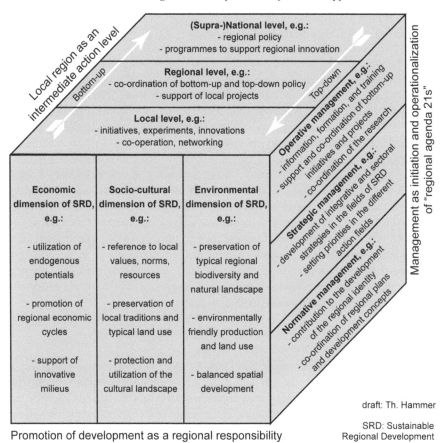

**Figure 2.2    The hexahedron of sustainable regional development**
*Source: Drafted by author*

must be able to initiate and maintain processes in the sense of regional Agenda 21s and to involve the actors. Together with the relevant actors the management of the protected area or the sponsoring organization must take a position on normative, strategic and operative questions. At the normative level the crucial question deals with what is to be achieved (normative management: what do we want to achieve?). The strategic level needs to clarify how these goals are to be achieved (strategic management: how shall we proceed?). And at the operative level the question of programmes and projects takes priority (operative management: which programmes and projects shall we initiate and how?).

From a theoretical standpoint it must be assumed that all of these conditions must be fulfilled if protected areas are actually to be employed as instruments of regional development. Thus before positive regional economic effects can be expected, important advance work must be done, though this does not guarantee that the objectives will actually be achieved.

**Regional Policy Support for the New Approaches to Regional Development in Connexion with Large Scale Protected Areas**

At all political levels, from the local to the national to the global, interest exists in establishing protected areas. The interests may change from one level to another, however, and may vary considerably from one actor to another. Whereas at the global level conservation is the paramount objective (for example preserving global biodiversity), the local level is more interested in goals related to regional development (regional actors expect especially socio-economic benefits from a protected area). At the national level, by contrast, objectives related to both conservation and regional development are usually important. On the one side the national actors want to achieve global and national conservation objectives and therefore support protected areas; on the other side there is an increasing tendency at the national level to employ the establishment and support of protected areas as an instrument of regional policy. The establishment of and national support for a protected area should help the regions to specialize, make better use of their natural and cultural resources, differentiate and distinguish themselves from other regions and thus improve their efforts at regional development. How this objective is handled in Europe will be explained in the following on the basis of some individual examples.

*Regional Policy Support for Protected Areas at the National Level*

As already indicated, a paradigm shift is taking place in national regional policy in the sense that innovative grassroots projects are receiving increasing support from 'higher up' – and this includes the establishment of protected regions. Despite this general trend there is, however, a great discrepancy in the national regional policies of European countries. The regional policy support for protected areas varies considerably from country to country. Whereas France was one of the pioneers and has for decades been considered a model of regional policy promotion of protected areas, in Switzerland this discussion only began a few years ago.

Since the mid-1960s France has supported regional nature parks as an instrument of regional policy (promoting marginal rural regions). Since the decree on regional nature parks proclaimed in 1967 by *Général de Gaulle* three objectives are paramount. Regional nature parks are supposed to (a) represent buffer areas for the large metropolises and serve as recreational areas, (b) promote socio-economic development in the marginal rural regions and (c) protect the regional flora and fauna and the landscapes and make better use of the natural and cultural resources. In the course of time further goals were added; the regional nature parks are supposed to play a role in regional planning and promote sustainable development.

With this relatively high level of support, 44 regional nature parks had been founded by 2005. They cover 12 per cent of the land area of France, and 3.1 million persons live in them. The average size of a nature park is around 150,000 ha (1,500 km$^2$) with an average of 70,500 inhabitants. In accordance with the relatively long tradition of nature parks and the relatively great significance for the urban and rural regions, a kind of nature park culture has developed that in the eyes of many people enhances the attractiveness and contributes to the regional image. With its regional nature parks France has also played a pioneer role in the development of relevant concepts, for which reason it is repeatedly portrayed as a model when the significance and promotion of protected areas are discussed.

In Switzerland, by contrast, after several years of intensive and highly controversial discussion a consensus is just beginning to emerge at the national level that protected areas should be promoted at least to a minimal degree by regional policies and that they should be employed as an instrument of regional policy. This has a lot to do with the fact that in Switzerland, more strongly than in France, regional policy is indirectly pursued in the context of sectoral policies (for example via agricultural policy) and direct regional policies were previously aimed one-sidedly at the regional infrastructure. Since for various reasons indirect and direct regional policies are scarcely able to fulfil the hoped for objectives, a change in regional policies is take placing towards increased support for innovative, promising grassroots regional initiatives and thus also for protected area initiatives.

For example, between 1997 and 2005 the *Impulsprogramm zur Unterstützung des Strukturwandels im ländlichen Raum – Regio Plus* [Impulse Programme for the Support of Structural Change in Rural Regions], which was modelled on the innovation programmes of the European Union, supported the planning and/or establishment of 12 protected area projects, which make up more than 10 per cent of all supported projects. These protected area projects are usually aimed at sustainable development and combine conservation and regional development. Significantly, the *Regio Plus* programme is financed by the State Secretariat for Economic Affairs, making it clear that the intention is to employ protected areas as instruments of regional development. Nevertheless, there is a recognizable tendency to charge the funds for the protected areas to the conservation budget, especially because the expected regional economic effects are rated by many actors as rather modest.

*Regional Policy Support for Protected Areas at the Supranational Level*

At the European level the European Fund for Regional Development plays an especially important role for general regional development. Since 1991 the special joint initiative *Liaison Entre Actions de Développement de l'Économie Rurale* (*LEADER*) is trying and promoting new approaches to rural regional development and regional governance whose purpose is to help rural actors make full use of the long-term development potential of their region through initiatives of their own. Innovative approaches and rural development projects are financed. The idea is to motivate the regions to develop and submit overall concepts combining the natural capital of the region with gentle types of use that add value to local products. In the third phase of the LEADER programme (LEADER+, 2000–2006) *local regions* or so-called *Local Action Groups* (*LAGs*) are to be motivated to develop and implement integrated strategies of sustainable development. These should foster partnerships and networks and be adapted to the local conditions. The quantitative focus of the supported projects, at 34 per cent of the projects, is on such projects that make good use of the regional, natural and cultural resources and sites. The other priority themes of the supported projects are also suitable for protected areas that intend to apply new approaches to regional development (see Table 2.2).

**Table 2.2        The priority themes of the projects in the European programme Leader+ (2000–2006)**

| | |
|---|---|
| Making the best use of natural and cultural resources, including enhancing the value of sites | 34% of the projects |
| Improving the quality of life in rural areas | 26% of the projects |
| Adding value to local products, in particular by facilitating access to markets for small production units via collective actions | 19% of the projects |
| The use of new know-how and new technologies to make products and services in rural areas more competitive | 11% of the projects |

*Source: European Communities 2003, 15*

In France a total of 140 local regions are supported by LEADER+ (2000–2006), 41 per cent of them with the priority theme 'development of natural and cultural resources' and 14 per cent with the priority theme 'development of local products'. Various regional nature parks are directly supported, for example the Regional Nature Parks Haut-Jura, La Brenne, Livradois-Forez, Morvan, Pilat and Volcans d'Auvergne. Other parks are backed indirectly, for instance through the support of LEADER+ for the activities of the *Local Action Groups* (*LAGs*) in which parks participate or through projects that affect actors within the parks.

According to the evaluation of the *European Communities* (European Communities 2006), the LEADER+ programme's strategic approach of supporting local initiatives has proved its worth. Consequently, the joint initiative LEADER+ is to be continued, and in the future all programmes of rural development are to contain a 'LEADER axis' for the preparation and implementation of local development strategies.

The trend towards regional policy support for innovative, integrative regional initiatives and for the newer approaches to regional development will thus presumably continue at both the national and the supranational level. For the establishment and development of protected areas the conditions thus look pretty good. Whether the protected areas will be able to reach their objectives and fulfil the expectations is not yet guaranteed.

**Challenges**

Bearing in mind the discussion of combining protected areas and regional development, various questions arise. They include the following:

- How are the new approaches to regional development being applied in the protected areas and with what consequences, for instance in the economic, ecological, social and cultural dimensions of sustainable development?
- What are the most important conditions and reasons for positive results? – What role do the institutional environment, the classification in a certain IUCN category, issues of acceptance and participation play?
- How important are the management of a protected area and the governance structures, and what experience was made with innovative concepts of management and governance structures?

- Is the spatial zonation of the protected areas, which provides for a graded intensity of protection and use according to the UNESCO Biosphere Reserve concept, an effective instrument for achieving the objectives?
- Do protected areas that pursue an integrated protection and development strategy make a better quantitative and/or qualitative contribution to regional development than protected areas that aim one-sidedly at protection or development?
- How are conflicts relating to objectives handled in protected areas? Does the solution of conflicts also contain a potential for development?
- How are win-win or win-win-win situations and synergy that extend beyond the dimensions of sustainable development created in the protected areas?
- To what extent does the experience in protected areas represent a model for overcoming problems in rural regions outside of the protected areas?
- What are the general lessons concerning the integration of conservation and development in protected areas that can be applied in research, practice and policy?

Not all of these questions can be dealt with comprehensively in this volume. Chapter 14 will summarize the information gained from the evaluation of the case studies.

## References

Butzin, B. (2000), 'Netzwerke, Kreative Milieus und Lernende Region – Perspektiven für die regionale Entwicklungsplanung?', *Zeitschrift für Wirtschaftsgeographie* 44: 3–4, 149–66.

Chape, S., S. Blyth, L. Fish, P. Fox and M. Spalding (compilers, 2003), *United Nations List of Protected Areas. IUCN, International Union for the Conservation of Nature and Natural Resources, and UNEP, United Nations Environment Programme.* (Gland, Switzerland and Cambridge, UK).

European Communities (2003), *Rural Development in the European Union.* (Luxembourg).

European Communities (2006), *New Perspectives for EU Rural Development.* (Luxembourg).

FPNRF (2004), *Les Parcs Naturels Régionaux – 37 ans d'histoire.* (FPNRF, Fédération des Parc Naturels Régionaux de France. Paris).

Finger-Stich, A.S. and K.B. Ghimire (1997), *Travail, culture et nature. Le développement local dans le contexte des parcs nationaux et naturels régionaux de France.* (L'Harmatten, Paris).

Fürst, D. (2001), 'Regionalmanagement als Instrument einer nachhaltigen Regionalentwicklung', in Behrens, H., P. Dehne, J. Kaether (eds): *Regionalmanagement – Der Weg zu einer nachhaltigen Regionalentwicklung.* 1–12. (Neubrandenburg).

Fürst, D. and J. Knieling (eds) (2002), *Regional Governance. New Modes of Self-Government in the European Community. Studies in Spatial Development 2.*

(ARL, Akademie für Raumforschung und Landesplanung, VSB-Verlagsservice, Braunschweig).

Hammer, T. (2001), 'Biosphärenreservate und regionale (Natur-)Parke. Neue Konzepte für die nachhaltige Regional- und Kulturlandschaftsentwicklung?', in: GAIA 10:4, 279–85.

Hammer, T. (ed.) (2003), *Grossschutzgebiete – Instrumente nachhaltiger Entwicklung.* (München).

Hammer, T. (2004), 'Schutzgebiete als Grundlagen lokal-regionaler Agenden nachhaltiger Entwicklung', in Gamerith, W., P. Messerli, P. Meusburger and H. Wanner (eds), *Alpenwelt – Gebirgswelten. Inseln, Brücken, Grenzen. Tagungsbericht und Abhandlungen des 54. Deutschen Geographentags Bern 2003.* 749–58. (Heidelberg and Bern).

IUCN (1994), *Guidelines for Protected Areas Management Categories.* IUCN, International Union for the Conservation of Nature. (Cambridge, UK and Gland, Switzerland).

Leimgruber, W. and T. Hammer (2002), 'Biosphere Reserves – Sustainable Development of Marginal Regions?', in Jussila, H., Majoral, R. and Cullen, B. (eds), *Sustainable Development and Geographical Space. Issus of Population, Environment, Globalization, and Education in Marginal Regions.* 129–44. (Ashgate, Aldershot, UK).

Mose, I. and N. Weixlbaumer (eds) (2002), *Naturschutz, Grossschutzgebiete und Regionalentwicklung.* (Sankt Augustin).

Scheff, J. (1999), *Lernende Regionen – Regionale Netzwerke als Antwort auf globale Herausforderungen.* (Linde, Wien).

UNESCO (1996), *The Seville Strategy for Biosphere Reserves. UNESCO, United Nations Educational, Scientific and Cultural Organisation.* (Paris).

# PART II
## Selected Case Studies

Chapter 3

# Biosphere Reserves: An Instrument for Sustainable Regional Development? The Case of Entlebuch, Switzerland

Thomas Hammer

UNESCO Biosphere Reserves are protected areas that are entitled to bear this label, having been nominated by the state in which they are located and passed a strict UNESCO evaluation procedure. In 2005 there were 482 protected areas in 102 countries recognized by UNESCO as biosphere reserves worldwide. Biosphere reserves have a commitment to conservation, and since the so-called Seville strategy of 1995 their objectives have broadened. Biosphere reserves are intended not only to be protected areas, but also to represent regional planning models, experimental regions of sustainable development and even model regions of sustainable development (see chapter 2).

But in what does sustainable regional development consist? What conditions need to be fulfilled before biosphere reserves can claim to pursue sustainable regional development? What authority needs to be granted to the management of a protected area so that it can carry out its responsibilities? What is the appropriate institutional embedding for the reserve? What can biosphere reserves effectively accomplish or what can they not accomplish? What lessons can be drawn from previous experience? What are the crucial challenges for the management of a protected area if it is to pursue sustainable regional development? And can biosphere reserves even fulfil their own high expectations? – These are the questions this chapter will attempt to answer, first in general and then specially in relation to the biosphere reserve Entlebuch. The biosphere reserve Entlebuch was chosen as an example because of its particularly high aspirations to be an instrument of sustainable regional development and because it is often considered to be a prime example even by specialists.

## Standards to be Met by Biosphere Reserves that are to Serve as an Instrument of Sustainable Development

In the research on biosphere reserves and sustainable regional development two perspectives can be roughly distinguished.

On the one hand, investigations can deal with the contribution biosphere reserves make to sustainable regional development. Accordingly their impact is analysed, for instance in the three dimensions of sustainable development. Often, however, only one dimension of sustainable development is investigated, for example only the ecological or only the economic impact of a protected area. Although sectoral analyses do not cover the entire spectrum of a sustainability analysis (if one proceeds from the sustainability concept of the United Nations), they are useful and necessary and their results can be included in an overall evaluation of sustainability. The aspiration of a biosphere reserve to contribute to sustainable development is usually relatively easy to fulfil (especially if only one dimension of sustainability is considered), because biosphere reserves, like other protected areas, usually contribute in some way or other to sustainable development.

On the other hand, research can be concerned with whether and how the biosphere reserve can be employed as an instrument of regional development and what impact this has. Underlying this is a highly ambitious objective: biosphere reserves should actually represent a tool for achieving the regional goals of sustainable development.

Particularly for this second concept (biosphere reserves as instruments of sustainable regional development) it is important to understand what is meant by sustainable regional development. If sustainable regional development is defined on the basis of the sustainability concept that goes back to the Conference on Environment and Development of the United Nations in 1992 in Rio de Janeiro, sustainable regional development consists:

- In a contribution to solving global problems and to achieving global objectives (see chapter 28 of Agenda 21)
- In a contribution to solving national problems and to achieving national objectives and
- In the solution of local-regional problems and achieving local-regional objectives.

Biosphere reserves can only be classified as instruments of sustainable regional development in this sense if they consciously and deliberately solve regional problems and attempt to achieve regional goals. We can distinguish priority objectives, that is ones aiming at the desired condition of the region as a whole. The concrete goals can normally be classified as belonging to one of the three dimensions of sustainable development, that is the economic, the socio-cultural, or the ecological dimension. The resulting first ideal task of the management of a protected area, as explained in chapter 2, is to promote development across the three dimensions of sustainable development in order to make the best use of the potentials with a minimum of conflicts. To be able to tackle this first goal the management must have the appropriate authority and funds. The communes themselves can form the regional level (for example in the form of an association of local authorities); in addition to their communal goals they must, however, pursue independent and joint regional objectives, so that active regional development becomes possible. The biosphere reserve thus becomes an intermediate level of action between the communes on the one hand and the higher levels (federal states, national level) on the other hand. The biosphere reserve initiates, motivates, arranges and co-ordinates projects between lower and higher levels and at the regional level and employs national programmes to achieve regional objectives.

As mentioned in chapter 2, the management of regional development encompasses the three classical dimensions of management, namely normative (definition of the objectives: 'what should the region achieve?'), strategic (planning: 'how should the region proceed?') and operative (initiation, co-ordination, implementation: 'what projects and programmes should be carried out?'). Naturally a biosphere reserve cannot assume all of these responsibilities. Rather, the management concentrates on operative management, which derives from normative and strategic management. It is important, however, to have regional objectives that were elaborated in accordance with concepts of sustainable development. These can exist in the form of regional models, Agenda 21s, charters, development concepts, or plans of action. The management of a protected area can then base its activities on these objectives and appraise the effected changes on their basis.

In view of the multi-level concept (global, national, regional, local level) and the equivalence of the three dimensions of sustainable development, management of protected areas signifies *regional action* that balances global, national, regional and local demands on the one hand and ecological, economic and socio-cultural objectives on the other hand. Precisely because the three dimensions of sustainable development are equivalent, it is important to formulate priority objectives, so that conflicts can be solved on the basis of higher regional goals. Creating synergies and win-win-win situations between the three dimensions likewise becomes a crucial challenge in view of the equivalence of the dimensions. The fundamental danger, however, is that when projects that do not result in a win-win-win situation are considered, preference will be given to ones that promise gains primarily in the economic and/or socio-cultural dimensions. These gains accrue more rapidly and are more obviously in the self-interest of the involved parties than those in the ecological dimension (and the losses in the ecological dimension are usually not recognizable until later).

## UNESCO Biosphere Reserves and Sustainable Regional Development

In order to allay the danger of giving priority to the economic and socio-cultural dimensions and neglecting the ecological dimension of sustainable development, it is advisable for the management of a protected area to follow the UNESCO Biosphere Reserve concept, even if no UNESCO certification is aspired to. As mentioned (see chapter 2), the biosphere reserves pursue ecological, socio-cultural and regional economic goals. These are to be achieved through zoning of the protected areas, distinguishing three zones according to the intensity of conservation and land use. Whereas in the core zone the priority is to protect the natural species diversity and allow natural processes to run their course, in the buffer or transitional zone the culturally influenced biodiversity that usually grew out of century-old traditional gentle types of human use is to be preserved and made better use of. Use and cultivation are thus not only tolerated in the transitional zone, but even desired or indispensable for achieving the objectives. In the development zone, in contrast, the focus is on environmentally friendly use of intensive land use systems ('integrative conservation') and on the persons cultivating them with their needs for work, homes, supplies, recreation and mobility. Here too the priority objective is to be a model for regional planning and an experimental area for sustainable development.

Such zoning or graded regulation of protection and use already exists in many large protected areas, usually at least in rudimentary form, because many large protected areas were first founded around already existing biotopes that were subject to strict conditions. The biosphere reserve concept, however, invites us to take a comprehensive view of spatial land use. It calls for a three stage regulation of protection and use, not the usual two stages. It provides an incentive to include the settlement areas and the intensively used areas in the protected area. The concept encourages setting spatial priorities for protection and use, combining fragmented core zones and embedding them in a buffer zone and including human forms of use in the management of the protected area. Fundamentally it encourages taking an inclusive view of the area. Spatial zones can be specified according to the priority objectives and in relation to the dimensions of sustainable development. This makes it easier to determine the extent to which the objectives are achieved.

If the management of a protected area is to pursue sustainable regional development, various conditions must be fulfilled. The hard conditions, that is those that have to be fulfilled in any case, include first of all the existence of a concept of the objectives of sustainable regional development. Second, the management of the protected area must be granted the necessary authority, and third it must have an appropriate level of funding and personnel.

Before the management can devote itself to sustainable development, a concept of the objectives of sustainable regional development must exist or be compiled. The management must know what actually is to be achieved. It must additionally possess the necessary authority or at least be able to take on appropriate responsibility. Likewise funding and personnel are indispensable if projects and programmes are to be initiated, promoted and supported.

Furthermore, there are soft conditions, that is ones that should exist or be aspired to. First, there should be an appropriate institutional embedding; second, the participating persons should have an appropriate self-image, and third, the participating persons should have appropriate management skills.

The institutional embedding of the protected area is usually very important. If the protected area belongs to a conservation or environmental agency, it is more difficult to pursue sustainable regional development than if regional actors from various sectors, the region as a whole, or the communes of the region are responsible for the protected area. In the same vein, the participating persons must have an appropriate self-image. The insight that conservation and sustainable regional development are fundamentally different fields of activity is crucial. Conservation can, however, be an important component of regional development, if it is defined as such. A well functioning interdisciplinary team and participating persons who have management skills are also crucial, and management encompasses much more than mere project management. The individual projects should instead be embedded in the region's overall development process in such a manner that a coherent process of sustainable development emerges. Process management thus becomes a great challenge, and it begins already with the development of visions and goals: What do we want? Why? And how can we achieve this? What concrete goals should we aspire to? How can the actors be incorporated into

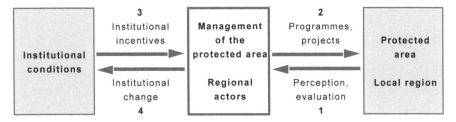

**Figure 3.1     Actor-centred model of the interaction between institutional
conditions and regional development**
*Source: Drafted by author*

the responsibility? What priority themes, programmes and projects should be developed, and how should the funds be employed? These are only some of the questions with which the management of a protected area must implicitly or explicitly come to terms.

What biosphere reserves can effectively achieve thus depends to a great extent on the objectives, the purpose, the funds and the institutional conditions. These are given facts, especially in a short-term perspective. Viewed in the medium and long range they are, however, modifiable. The management of a protected area can try to effect changes and, for example, utilize the institutional environment for its concerns (including national and international programmes). In accordance with the institutional control approach (see Figure 3.1) the management of a protected area can act in four areas:

- First, in the perception, evaluation and definition of objectives (arrow 1 in Figure 3.1): the management of the protected area or the persons responsible for sustainable regional development can influence the perception, evaluation and definition of objectives. Especially the experience with protected areas shows that the normative superstructure can change relatively rapidly in the course of time.
- Second, in direct action towards sustainable regional development (arrow 2): the management of a protected area or the regional actors are relatively free at the operative level and, depending on the conditions, at the strategic level as well, and can set priorities, for example when they initiate programmes and projects. They can support projects that are beneficial only to conservation or the regional economy or ones that create win-win or win-win-win situations across the dimensions.
- Third, in utilizing the institutional conditions (arrow 3): the management of the protected areas or the regional actors can utilize the institutional conditions for their concerns to a greater or lesser extent. There are national and international policies and programmes offering incentives to the regions (at the European level LEADER+, INTERREG III and Natura 2000; at the international level the UNESCO labels; at the national level programmes for promoting regional innovation), which can be utilized for sustainable regional development.
- Fourth, in influencing the institutional conditions (arrow 4): the managers of a protected area and the regional actors themselves can work towards changes in the institutional environment, at least at the regional and local level.

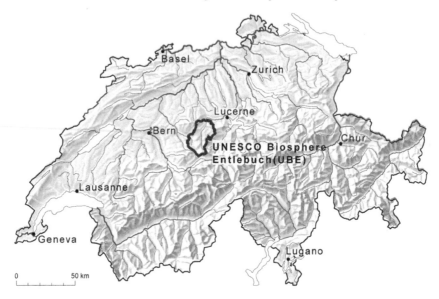

**Figure 3.2     Location of the UNESCO Biosphere Entlebuch within
             Switzerland**
*Source: Drafted by author*

The development or revision of regional models and concepts of regional
(landscape) development, adaptations to regional planning, or changes in
the purpose of the protected area can definitely be influenced. Some of the
institutional conditions can also be modified regionally.

According to this general model acting towards sustainable regional development
means to advocate the drafting of a model, to initiate goal oriented programmes
and projects, to utilize the institutional conditions and to work towards changing
especially the regional and local conditions in favour of sustainable regional
development. This approach assumes that the existing conditions, from the local to
the international level, represent both opportunities for and obstacles and limits to
regional action. Especially the national and international conditions must be viewed
as givens; the opportunities resulting from the national and international conditions
can be utilized nevertheless, depending on the regional potentials. For this purpose
regional analyses of the situation and environment are indispensable.

An example of a biosphere reserve that is employed as an instrument of sustainable
regional development is the UNESCO Biosphere Reserve Entlebuch.

**The UNESCO Biosphere Entlebuch**

*Origin, Objectives, Zonation and Institutional Embedding of the Protected Area*

The UNESCO Biosphere Entlebuch (this is the official name of the biosphere reserve
Entlebuch) lies in the Alpine foothills in the heart of Switzerland, in the marginal

**Table 3.1     Structural data on the UNESCO Biosphere Entlebuch**

| | |
|---|---|
| Area | • 39,500 ha (approx. 1% of the area of Switzerland) |
| Inhabitants | • approx. 17,000 (around 0.25% of the population of Switzerland) |
| Protected areas of | • 12,100 ha landscapes of national importance |
| national importance | • 10,400 ha mire landscapes |
| (partially overlapping) | • 1,765 ha fen |
| | • 170 ha bog |
| Land use | • forest 43% |
| | • agriculture 30% |
| | • alpine pastures 18% |
| | • unproductive 7% |
| | • settlement 2% |
| Employment (1995) | • 37% in the primary sector |
| | • 24% in the secondary sector |
| | • 39% in the tertiary sector |
| Zoning | • a. core zone 8% (3,301 ha) |
| | • b. buffer or transition zone 42% (16,358 ha) |
| | • c. development zone 50% (20,000 ha) |
| | An extension of a. is intended. |
| Swiss requirements | • area: 20,000 to 100,000 ha |
| for biosphere reserves | • core zone: greater than 3% (or 5% if noncontiguous) |
| | • buffer or transition zone: greater than 10% |
| | • core and buffer zone together not under 20% |
| | • development zone: 50–80% |

rural region between the cities of Lucerne and Bern (see Figure 3.2). With its almost 40,000 ha it encompasses about 1 per cent of the area of Switzerland and with its approx. 17,000 inhabitants around 0.25 per cent of the national population (see Table 3.1). On the basis of its graded zonation of protection and use the UNESCO Biosphere can be classified in the IUCN categories four (Habitat/Species Management Area) and five (Protected Landscape). The protection of the core and buffer zones is guaranteed by national law and subsidiarily by cantonal law (see Figure 3.3).

Beginning in 1997, the idea to establish a biosphere reserve, which arose out of initiatives of the regional planning association (association of local authorities), received support from many sides. From the beginning it was thought of as an instrument of regional development. The communes, the Fonds Landschaft Schweiz (FLS), the national *Programm zur Unterstützung des Strukturwandels im ländlichen Raum* [Action programme to overcome structural changes and promote co-operation in rural areas = *Regio Plus*] and further institutions supported the idea and enabled the preparatory work for an application for recognition as a biosphere reserve (see Table 3.2). In the course of this preliminary work the rough draft of a 'Biosphere Reserve Entlebuch' emerged beginning in 1999 (see Regionalmanagement BRE 2002). It was developed participatively and can de facto, together with the model of the board of the regional planning association of 2002, be interpreted as the first model of sustainable regional development in Entlebuch. The priority objective reads as follows:

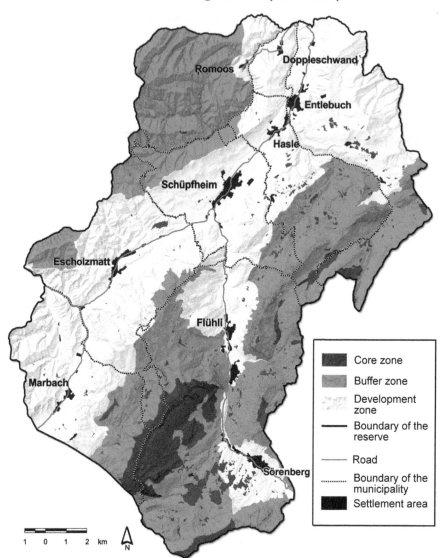

**Figure 3.3     Zonation of the UNESCO Biosphere Entlebuch**
*Source: Drafted by author*

The UNESCO biosphere reserve Entlebuch ... is developing into a model region with regard to preservation of the natural and cultural landscape, sustainable regional development, participation of the population, co-operation and management (Vorstand des Regionalplanungsverbandes UBE 2002).

The rough draft explains how this is to be attempted. It distinguishes goals related to *preservation*, *development*, *participation* and *co-ordination and co-operation*:

**Table 3.2    Data on the formation of the UNESCO Biosphere Entlebuch**

| | |
|---|---|
| **1987** | • The popular initiative calling for constitutional conservation of the Swiss mire landscapes (so-called Rothenturm Initiative) is adopted. |
| **til 1996** | • A directive plan 'Moor landscapes' is formulated and the biotope and landscape conservation called for in the Rothenturm Initiative implemented. |
| **1997** | • Start of the project 'Habitat Entlebuch' supported by the regional planning association Entlebuch and partially financed by the Fonds Landschaft Schweiz. |
| | • Information gathering on the establishment of a biosphere reserve starts. |
| | • The regional management is organized. |
| **1998** | • The regional management of the 'Project Biosphere Reserve Entlebuch' (1998–2001) starts operating. |
| | • Broad support for the project, *inter alia* from RegioPlus (1999–2001). Total costs 1999–2001 approx. 1.8 million SFR. |
| | • The regional planning association sets up a project committee. |
| | • Various working groups and the Förderverein (association of the friends of Entlebuch) are founded. |
| **1999** | • The rough draft is formulated with the participation of the local people: model, zonation, educational concept, research strategy, goals for a review of the results, PR concept. |
| | • Further information, communication and co-operation with local, regional, national and international partners. |
| | • Brands are established: 'Tourism Entlebuch', 'Products from Entlebuch'. |
| **2000** | • At communal meetings in September 2000 a vast majority of the inhabitants approve financial support for the biosphere reserve amounting to SFR 4.- per capita and year for 10 years (on average 94% votes in favour). |
| | • The biosphere reserve is registered with the Canton of Lucerne, the Federal Environmental Agency (Bundesamt für Umwelt, BAFU) and the Federal Government, all of which promise (further) support. |
| **2001** | • The Federal Government forwards the registration to UNESCO, which recognizes Entlebuch in September as the first Swiss biosphere reserve in accordance with the Seville strategy. |
| | • The new sponsoring organization is established. |
| | • The brand name 'echt entlebuch biosphärenreservat' ('true entlebuch biosphere reserve') is introduced. |
| **2002** | • The Biosphere Reserve Entlebuch is officially admitted to the UNESCO world network of biosphere reserves. |
| | • The rough draft, which had been continuously refined since 1999, is published (Regionalmanagement BRE 2002). |
| **2003** | • The Biosphere Reserve Entlebuch receives the new official designation 'UNESCO Biosphere Entlebuch – Lucerne, Switzerland' reflecting the expanding marketing strategy. |

• The total of five *preservation goals* refer to the natural, semi-natural and culturally influenced landscape, to the human and nonhuman biological habitats, agriculture and forestry, economic and cultural diversity

- The nine *development goals* focus on creating opportunities for jobs and job training (especially for young people) and for earning additional income, developing co-operation between various sectors of the economy, promoting high quality tourism and developing private and public transportation in an environmentally sensitive manner
- The two *participation goals* aim at sensitizing the regional population to relevant developments within and outside of the region and at designing the biosphere reserve together with the population and
- The ten *co-ordination and co-operation goals* aim at close collaboration with various partners and the population, at co-ordinating activities and involving the partners and the population in achieving the objectives of the biosphere reserve, for instance in continuing education, sustainability monitoring, environmental observation, biosphere GIS and catering to guests.

In the biosphere reserve Entlebuch the most important conditions for pursuing sustainable regional development by means of a protected area concept – in this case the UNESCO biosphere reserve concept – exist. The involved actors participated in the formulation of the objectives; the biosphere management was established along the lines of regional management; institutional incentives (in particular specific programmes promoting regional innovation) were utilized; the biosphere management was granted the requisite authority; the necessary funds were made available and institutional adaptations were made.

Against the background of the above explained model of institutional regulation and the conditions for pursuing sustainable regional development two aspects should be especially mentioned:

- First, the institutional regulation of participation and the institutional embedding of the biosphere reserve: the biosphere management is primarily an operative organ of the communes and thus of the regional planning association (and not of a higher level of the administration). The biosphere reserve belongs to the political communes and thus the inhabitants, who themselves determine the communal policies to a great extent. The inhabitants, the local and regional actors and groups of actors are active in various forums. These forums represent important sources of new projects. The biosphere management supports the forums and the co-ordination committee that is being developed co-ordinates the activities between the forums. An active participation management guarantees that the inhabitants and the actors will continue to be involved.
- Second, the *access to sustainable regional development*: in order to achieve the above explained goals, *regional cycles* are being promoted. According to the rough draft this means that in the environmental dimension of sustainable development *flows of matter* are to be on as small a scale as possible and *cycles of matter* to be closed. In the social dimension it means promoting local and regional *action chains* through co-operation, and in the economic dimension developing and closing regional *value-added chains* and thus helping to supply the inhabitants with their basic needs.

An interim stocktaking after three years of preparation and four years of UNESCO recognition shows that the actors succeeded in developing concepts for sustainable regional development in Entlebuch, in using the institutional conditions in a goal oriented manner, in bringing about an adequate institutional embedding for the biosphere reserve, in reconciling the UNESCO biosphere reserve concept with endogenous ideas of sustainable regional development, in employing the biosphere reserve as an instrument of sustainable regional development, in initiating a great variety of projects and in keeping alive the development process as a whole. Altogether the conditions that were created allow sustainable regional development to be pursued with the biosphere reserve.

*Impact of the Biosphere Reserve in the Dimensions of Sustainable Regional Development*

Whether all of the goals have been achieved with the protected area is difficult to assess. Therefore impact hypotheses were formulated.

*Hypothesis 1*   Many qualitative goals were achieved, at least partially. Examples are the development of co-operation between the various sectors of the economy or the sensitization of the inhabitants to relevant developments within the region. Others are that the biosphere reserve was designed together with the inhabitants. Close collaboration with various partners and the inhabitants has developed and is being co-ordinated. The inhabitants have been involved in achieving the objectives, and a regional geographic information system (GIS) has been developed. Parts of the population and the regional actors were successfully involved. Indications of this are the participation in the thematic forums, in the activities and in the many projects and the generally high level of acceptance of the biosphere reserve among the inhabitants meanwhile.

*Hypothesis 2*   No concrete data exist regarding quantitative goals such as creating opportunities for earning additional income, for job training or jobs and for value-added chains. it can be concluded from the obvious increase in nature oriented trips and in excursions in general, the increasing sale of products with a regional label and the attractiveness of the new offers (including Kneipp facilities, Seelensteg, Hochseilpark) that the biosphere reserve has a positive impact.

*Hypothesis 3*   The qualitative improvements in the range of tourist activities also point to a positive impact. On the one hand, new special activities directly related to the biosphere reserve are being marketed. On the other hand, the biosphere management itself is setting up a tourism office to create high quality regional activities and co-ordinate the tourist activities regionally.

*Hypothesis 4*   In contrast, there has been hardly any observable impact so far on other important goals, such as a more environmentally sensitive development of private and public transport or the preservation of biological habitats. Although activities exist in these areas, their impact is difficult to assess.

*Hypothesis 5*  On the whole the biosphere reserve represents an instrument of sustainable regional development and has a basically positive influence on regional development. The results are, however, gradual in nature. The basic problems of Entlebuch can only be alleviated, not solved, in view of the endogenous and exogenous conditions. Accordingly crucial is the cultural dimension of sustainable regional development. With the establishment and development of the biosphere reserve, Entlebuch adopted a forward strategy. Within only a few years it was transformed from the onetime 'poorhouse of Switzerland' to a self-confident model rural region that knows how to help itself. Here lies a considerable potential for development, especially for the regional economy.

The most important impact is possibly in the cultural dimension, namely reinforcing the regional self-confidence, which is an important basis for any development. Qualitatively much has changed for the better in Entlebuch in seven years in all three dimensions of sustainable development. Quantitative statements on the impact of the establishment and operation of the biosphere reserve can only be made about certain aspects, however. In particular, a comprehensive regional economic analysis is lacking.

## *Lessons to be Learned from the Biosphere Reserve Entlebuch*

With reference to the actor-centred model of the interaction between institutional conditions and regional development (see Figure 3.1) at least the following five lessons can be formulated based on the example of Entlebuch:

Firstly, the *quality of the process management* is of crucial importance. It appears to be a special skill to be able to trigger and maintain self-sustaining processes of sustainable regional development and keep the process as such in motion. Here we can distinguish four overlapping areas of management:

- *Participation management* consists in involving the inhabitants and actors not only in the normative and strategic, but also in the operative management of regional development, and this participation must extend to accepting responsibility for projects, if self-supporting processes are to be initiated. As soon as the actors themselves take responsibility for carrying out projects, the function of the protected area management changes towards support and counselling.
- *Co-operation management* means bringing together partners from various sectors of the economy so that ideas for innovative projects have a chance to develop in the first place. In accordance with Entlebuch's approach to sustainable regional development (see above), namely creating regional cycles of matter, action chains and value-added chains, the regional action chains represent the basis for the creation of regional cycles of matter and of value-added chains. Only through cross-sectoral action can regional cycles of matter and value-added chains develop.
- *Co-ordination management* comprises the co-ordination of ideas, activities, working groups and projects commensurate with the objectives of regional development. Sustainable regional development is a complex process that

requires co-ordination, because otherwise there is a danger that the individual parts will not result in a coherent whole.

- *Conflict regulation management* represents the fourth crucial area of management. Behind unutilized potentials, lack of initiative or abandoned use there often lie conflicts of various types. Instead of evading conflicts, the important point lies in solving them. Negotiation can bring differing interests into agreement and create synergies so that instead of deadlock, in which all initiatives are blocked, win-win or win-win-win situations can arise. The creation of such win-win situations is possibly the most important challenge to management.

Secondly, the *appropriate resources* must be made available to the management of a protected area, both financial and personnel. Sustainable regional development does not result by and of itself. Above and beyond the actual financing of the project the process management must be financed. The problem here is that the economic profits that the management of a protected area will earn cannot be calculated in advance, and even in retrospect they can only be calculated indirectly. In view of the scarce funds, financial backers have trouble mustering insight into the necessity of financing management services outside of the actual project management. And yet advance services are required if a subsequent regional economic impact is to be achieved.

Thirdly, the *socio-cultural dimension of sustainable development* represents the crucial link between the ecological and the economic dimensions. The motives for establishing a protected area are usually explained by outside actors in terms of the environment and/or economy, but the socio-cultural dimension plays several roles, particularly for internal actors. The priority objectives such as (a) preservation of the (cultural) landscape, the habitat for humans, the flora and fauna and the 'traditional' forms of use, (b) increasing the attractiveness and opening perspectives for the region and the young inhabitants, (c) improving the quality of life and preventing out-migration, and so on, are usually classified as belonging to the socio-cultural dimension of sustainable development. Moreover, sustainable regional development in the economic and ecological dimensions is created by the efforts in the socio-cultural dimension. The action chains are a prerequisite before cycles of matter or value-added chains can even develop. And not least, the impact on the socio-cultural dimension is most obvious and possibly the most important one: that a vision for the future emerges, a prospective regional identity, a belief in the region's own potentials and abilities. This again is an important basis for maintaining the endogenous process of creating chains of action, added value and matter.

Fourth, a certain, not readily quantifiable *size* is important for the initiation of a process of sustainable regional development. The size refers here to the number of inhabitants, the groups of actors, the sectors of the economy, the variety of types of production and habitats. If regional action chains, cycles of matter and economic cycles are to be created and zonation is to be done in the sense of the UNESCO biosphere reserves, a certain endogenous variety and differentiation are indispensable. Otherwise habitats, sectors of the economy and actors cannot be linked up into networks that give rise to novel products across sectoral boundaries.

And fifth, the *existence and the multidimensional utilization of regional peculiarities and common interests* is an important condition for initiating a process of sustainable regional development. Where there are no regional peculiarities and common interests it should prove difficult to elaborate common goals, to bring the actors together and to create regional chains of action and added value. In Entlebuch the shared history, the marginal location, the spatial unity, the moor landscapes and the protection of these landscapes form unifying elements that simultaneously represent potentials for regional development. The important thing is that making use of the potentials not be understood one-sidedly as purely economic use. For example, the material 'wood' or the moor landscapes have not only an economic value, but also an ecological and socio-cultural value. Only if this value is viewed comprehensively and comprehensive use is made of it, will its significance for sustainable regional development become apparent. Since expressing ecological and socio-cultural values in monetary terms requires a great effort, it is accordingly difficult to assess in economic terms the achievements of sustainable regional development as a whole, that is in the three dimensions of sustainable development.

## Conclusions

The example of the UNESCO Biosphere Entlebuch shows that biosphere reserves can be employed as instruments of sustainable regional development, if the appropriate conditions are filled. These include, in addition to the usual expectations towards concepts of sustainable regional development, granting the appropriate authority to the protected area management, making funds and personnel available and an appropriate institutional embedding of the protected area.

An essential, if not the essential, function is process management (project management alone is not sufficient!). This includes participation, co-operation, co-ordination and conflict regulation management. Only thus will an environment emerge in which the inhabitants and the actors can be involved in sustainable regional development and can act accordingly, that is initiate and carry out projects that contribute to achieving sustainable regional development. In addition to process management it is just as important to work towards an institutional embedding and funding commensurate with the task and to take advantage of the institutional conditions. All in all, it is important that the management of the protected area does not see itself as an administrative agency, but as an assertive management centre.

Sustainable regional development in accordance with the three dimensions of sustainable development can only be assessed appropriately on an interdisciplinary or transdisciplinary basis. The desired regional chains of action, matter and added value usually have an impact on all three dimensions, for which reason not only the economic profitability, but also the ecological and the socio-cultural profitability must be considered, otherwise the result is a one-sided evaluation that does not do justice to the issue at hand. Not least, the inhabitants and the actors should be involved in the evaluation (transdisciplinary aspect). What measures of regional development mean precisely in the socio-cultural dimension can only be determined by involving the inhabitants. And this makes social science research just as important as biological-ecological and economic research.

The question poses itself how positive experience resulting from combining biosphere reserves and sustainable regional development can be better propagated. A brief answer: in the 482 UNESCO biosphere reserves worldwide an immense stock of experiential knowledge exists of which much better use could be made. Exchange of experience and co-operation with other biosphere reserves across national boundaries can be very effective, as not only the experience in Entlebuch shows. Finally, sustainable regional development is based on a constant learning process and this only takes place if the actors recognize and take advantage of opportunities for the future across boundaries of all kinds. If those who are searching and those who are learning strive together for regional goals, form networks, co-operate and, so to speak, form a learning region, sustainable regional development can take place.

## References

Deutsches MAB-Nationalkomitee (ed.) (2004), *Voller Leben. UNESCO-Biosphärenreservate – Modellregionen für eine Nachhaltige Entwicklung.* (Berlin, Heidelberg).

Erdmann, K.-H. and Niedeggen, B. (2003), 'Biosphärenreservate in Deutschland – Lernräume einer nachhaltigen regionalen Entwicklung', in Hammer, T. (ed.), *Grossschutzgebiete – Instrumente nachhaltiger Entwicklung.* (München), 97–119.

Hammer, T. (2002), 'Das Biosphärenreservat-Konzept als Instrument nachhaltiger Regionalentwicklung? – Beispiel Entlebuch, Schweiz', in Mose, I. and Weixlbaumer, N. (eds), *Naturschutz: Grossschutzgebiete und Regionalentwicklung.* Reihe Naturschutz und Freizeitgesellschaft, Band 5. (Aachen), 111–35.

Hammer, T. (2003), 'Mensch – Natur – Landschaft. Exkursionen im UNESCO-Biosphärenreservat Entlebuch'. *Geographica Bernensia*, vol. 14. (Bern).

Hammer, T. (2005), 'UNESCO-Biosphärenreservate als Instrumente nachhaltiger Regionalentwicklung in Berggebieten – Am Beispiel der Biosphäre Entlebuch (Schweiz)', in Erdmann, K.-H. and H.-R. Bork (eds), *Zukunftsfaktor Natur – Blickpunkt Berge und Gebirge.* (BfN, Bundesamt für Naturschutz, Bonn-Bad Godesberg), 109–24.

Neubert, F. and Steinmetz, E. (2002), 'Die regionale Biosphärenreservats-Agenda 21. Chancen und Stand der Umsetzung der Agenda 21 in den Biosphärenreservaten Deutschlands'. *Deutsches MAB-Nationalkomitee, MAB-Mitteilungen* 46. (Bonn).

Regionalmanagement BRE (2002), 'Das Modell Entlebuch. Grobkonzept Biosphärenreservat Entlebuch', *Berichte aus der Region Entlebuch*, 2. (Schüpfheim).

Schmid, A. (2004a), 'UNESCO Biosphäre Entlebuch – Modell für eine nachhaltige Regionalentwicklung? Konzept Zielerreichungskontrolle', *Berichte aus der Region Entlebuch*, 3. (Schüpfheim).

Schmid, A. (2004b), 'UNESCO Biosphäre Entlebuch – Modell für eine nachhaltige Regionalentwicklung? Konzept Zielerreichungskontrolle', in *Geographica Helvetica* 59:2, 144–53.

Thierstein, A. and Walser, M. (2000), *Die nachhaltige Region – Ein Handlungsmodell.* (Bern, Stuttgart, Wien).

UNESCO Biosphäre Entlebuch, http://www.biosphaere.ch.

UNESCO (1996), *The Seville Strategy for Biosphere Reserves.* UNESCO, United Nations Educational, Scientific and Cultural Organisation. (Paris).

Vorstand des Regionalplanungsverbandes UBE (2002), *Leitbild der UNESCO Biosphäre Entlebuch.* http://www.biosphaere.ch.

Wallner, A. (2005), *Biosphärenreservate aus der Sicht der Lokalbevölkerung. Schweiz und Ukraine im Vergleich.* (Birmensdorf).

Chapter 4

# Can Tourism Promote Regional Development in Protected Areas? Case Studies from the Biosphere Reserves Slovensky Kras and Polana, Slovakia

Birgit Nolte

## Introduction

During the last decades, profound changes occurred in Central and Eastern Europe. The transformation processes have consequences in every part of life. For nature conservation this means an increased pressure on the protected areas, recently. In the socialist time, development was concentrated on economic centres while peripheral areas and especially border regions were left nearly untouched. As a result, many areas could conserve their high biodiversity, for example bear, wolf and beaver are still found in Central and Eastern Europe whereas in Western Europe these species are nearly extinct. However, this richness is increasingly threatened by the transformation process from a socialist planning economy to a western market economy. The most serious impacts on the environment are caused by unsustainable exploitation, pollution and land-use changes throughout Central and Eastern Europe. Moreover, tourism activities became increasingly a threat for biodiversity, in both a quantitative and qualitative sense.

Therefore, new protected areas were established and new strategies are needed to manage conflicting interests in the protected area.

In this context biosphere reserves, as internationally dedicated in the framework of the 'man and biosphere' programme of the UNESCO, provide opportunities and solutions. The objective of a biosphere reserve is to achieve a sustainable balance between the often conflicting goals of conserving biological diversity and promoting human development, while maintaining associated cultural values. Since the Seville strategy from 1995 this integrated approach has been applied within biosphere reserves. They are 'spaces for reconciling people and nature' (UNESCO 2002, 6).

The situation in Slovakia is characterized as typical for the countries joining the EU in 2004. The Slovakian government made great efforts to harmonize their legal framework to EU standards, especially in the field of nature conservation and established the ecological network NATURA 2000.

Slovakia's landscape, situated in the central eastern part of Europe, is very hilly and covered with forest. More than 40 per cent of the territory is above 300 m. They have a high proportion of protected areas because biological diversity is especially high. In addition these areas are also attractive for tourist activities, for example mountains are suitable for winter sports and hiking in summertime. Tourism in Slovakia is a growing market but is not yet seen as the main opportunity to develop the country. There are many untapped potentials for tourism development, primarily the attractive landscape.

Consequently, the two biosphere reserves studied demonstrate the recent situation of regional development through tourism coping with the interests of nature conservation. Three main sections provide the background for the case studies Slovensky Kras and Polana: firstly, the national framework for regional development, secondly, tourism development in Slovakia and finally nature conservation in Slovakia.

## National Framework of Regional Development

Slovakia has 5.4 million inhabitants and consequently a low population density. Only in the capital city Bratislava (600,000 inhabitants), the population is continuously growing because of its positive economic development (Statistical Office of the Slovak Republic 2001). Similar to other countries in Central and Eastern Europe there are two dominating regional characteristics:

1. West-east cascade from developed regions to periphery
2. Dominant role of capital region Bratislava, because of its western location and the proximity to Vienna (26 per cent share of Slovak GDP is produced in Bratislava region by only 12 per cent share on total population) (see Bucek 2000, 403).

Recent economic patterns are affected by the previous period of the directive planning system. For instance, the employment was the predominant task for macro-economy with the positive effects of low unemployment and few regional differences in economic development. But this was accompanied by low economic efficiency, huge redistribution between branches and regions, low income per inhabitant, degradation of environment and so on. The former regional policy with central planning solved regional differences by mobility orientated strategy, for example centrally planed location of enterprises.

The Velvet Revolution in the beginning of the 1990s generated dramatic changes with also regional implications. For instance, the regional disparities in unemployment grew considerably, but also other regional indicators like GDP, infrastructure endowment and productivity show remarkable disparities. The regional differences can be summarized as follows (see Bucek 2000):

• Bratislava is a core region of Slovakia's economy followed by Kosice (240,000 inhabitants) and the districts with developed and urbanized centres. Bratislava connects Slovakia with the western European countries and their markets. This is the reason why Bratislava as a centre of growth has such a dominant position.

- Transformational changes in macro- and microeconomics have regional variations, e.g. the changes in share of primary, secondary and tertiary sector, service sector development, increase of foreign and international enterprises and growing of the small and medium sized enterprises.
- The unequal allocation of hard and soft infrastructure (telecommunication, transportation, energy, etc.) caused regional disparities and competitiveness.
- There are regional differences in the allocation of research and science, technological development and the qualifications of the labour force.

To solve national economic problems the national policy is to promote centres of economic growth like Bratislava or for heavy industry like the Kosice region. There are attempts to change the former centralized planning and centre dominated policy, for example support for foreign investments in the Bratislava region. But regional disparities are growing in Slovakia despite the fact that a wide range of regional policy tools has been developed with the aim to create the necessary (soft) infrastructure for business and regional development.

For the future the dominant role of Bratislava will not be diminished because of its important position and good location in Slovakia. It is very likely that the regional disparities will continue to grow over the next few years.

> Reorientation of Slovak foreign trade from East to developed west market is going to dominate the whole regional development of the country. It will be the most important development impulse in the future (Bucek 2000, 405).

The situation of the people in the periphery can be described as 'forgotten', many still waiting for national assistance and state support. But like examples of European peripheries have demonstrated, local inhabitants have to concentrate on their own potentials and natural resources. In Slovakia in many cases, peripheral areas have high natural potential. To use these resources in a sustainable way could be the solution for regional development problems, such as high unemployment rates, low qualifications and low living standards. In short: a place to live for. The combination of tourism development and nature protection could be a promising way to reach this target.

## Tourism in Slovakia

To understand the tourism in Slovakia it is useful to describe the tourism development in Central and Eastern Europe in general. All of former Soviet bloc satellite states have experienced dramatic changes in tourism due to the transformation processes since 1989. Firstly, the domestic demand collapsed in the beginning of the 1990s. Secondly, traditional international markets were lost. Finally, western European markets had to be discovered. In consequence, a wholesale privatization of tourism provision took place due to a concerted attempt to create a market economy in a short

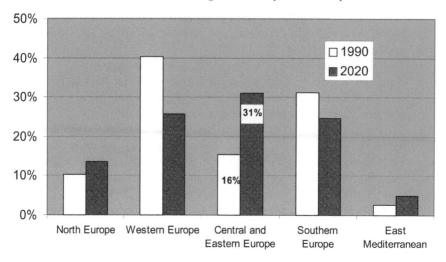

**Figure 4.1     Share of tourist arrivals in Europe by regions 1990 and 2020**
*Source: Drafted by author after BMWA 2003, 34*

time period. Additionally, the tourists' demands and behaviour changed in Central Europe caused mainly by international trends, for example individualism, hedonism and more frequent, shorter holidays. It is notable that against this background the tourism development in Central and Eastern Europe has grown constantly since the 1990s. Compared to other European regions, today the countries of Central and Eastern Europe have only a small share of tourist arrivals in Europe. However, an extreme growth in tourism is predicted for 2020 with the result that it will represent one third of all tourist arrivals in Europe (see Figure 4.1).

In Slovakia, tourism contributes only marginally to the national economy despite having a long tradition in spa tourism. New establishments and investments in this tourism sector answer the global trend in health tourism. Beside this, winter sport and nature orientated activities like hiking and biking are popular for Slovak's tourists. However, in general despite Slovakia's small size, there is no concentration on a specific tourism. Tourism development is not equally spread over Slovakia. Mariot describes three tourism zones characterized by tourism intensity (see Mariot 2003, 26). These zones are:

1. Major tourism zone: mountains in the north of Slovakia with the highest peaks and alpine landscape (for example High Tatras, Mala Fatra), regions that are very attractive for winter sport.
2. Intermediate tourism zone: regions adjacent to major tourism zone with high potentials for tourism activities and more intensive tourism development in future (for example large protected areas with attractive landscape).
3. Peripheral tourism zone: regions mostly situated in the south of Slovakia with a tourism highly dependant on the season (for example small villages with thermal springs).

In comparison with neighbouring countries tourism development in Slovakia is at a very low level. Especially in relation to tourism development in the Czech Republic, Slovakia's efforts to encourage tourist activities are weak. Despite this, tourism has been growing since 1989, but the rates of the last years were not always positive. However, since 2001 tourism has been continuously growing. In 2001 over 5 million overnight stays were estimated whereas in the Czech Republic the figures are much higher with 22 million overnight stays. The rate of foreign tourists is an important economic factor for Slovakia's tourism. Typical for countries in Central and Eastern Europe, over half of the Slovak's tourists came from abroad. However in the Czech Republic, between 1995 and 2001, this rate was always over 50 per cent (more than 60 per cent of the overnight stays in 2001 were made by foreigners in the Czech Republic). In contrast, the rates in Slovakia in the same period were always slightly under 50 per cent; however, in 2001 exactly 50 per cent of overnight stays were made by foreigners in Slovakia. In the Czech Republic 67 per cent of incoming tourists came from EU15[1] in 2002 compared to only 25 per cent in Slovakia (see Slowakische Zentrale für Tourismus 2004). In Slovakia however, this percentage grew about 13.3 per cent between 2001 and 2002 (see European Commission 2003, 32).

The role of protected areas in Slovakian tourism is not clearly defined. On the one hand, many nature conservationists emphasize the strong need to protect nature against tourism activities. On the other hand, the Slovakian tourist board considers that nature is one of the most important tourism factors in Slovakia, due to the strong linkage between tourism development and the natural landscape (see Slowakische Zentrale für Tourismus 2004). Consequently, Mariot stated that protected areas and especially national parks are the most important tourist destinations in Slovakia (see Mariot 2003, 19). The example of the High Tatra National Park demonstrates its importance as a tourist destination in Slovakia. The High Tatras became one of the first national parks in Slovakia in 1948. Its territory in the north of Slovakia covers more than 741 km². The so called smallest European mountain range continues on the Polish side where large protected areas are still not established. In the whole area we can find a long history of tourist arrivals and also for nature protection. With 1.6 million overnight stays in the year 2000 the High Tatra has the leading position in Slovakian tourism. Tourist activities are affected by the rich natural landscape and the cultural heritage. Especially winter tourism, hiking, and biking are the main activities as is typical for Slovak's tourist destinations. During the socialist era, tourist intensity was higher than today: At the end of the 1980s more than 5 million visitors were estimated in the High Tatras. According to the political importance of tourism as a social function, organization and costs as well as tourist expectations and behaviour was very different in this time. Nearly all tourist journeys were arranged by the trade union organizations. The costs normally included travel expenses as well as accommodation. The prices were comparatively low, as was the standard of the tourist facilities. The tourist expectations and behaviour have changed because of the changing society and related values.

---

1   EU15: the European Union consisting of 15 member states until 2004.

Today, protected areas can be used as a tool for solving economic problems by integrating tourism into nature protection, especially in Slovakia's rural areas where successful regional development is urgently needed.

## Nature Protection in Slovakia

Nowadays the national concept of nature protection is still connected with the past socialist system. Under the former socialist system, industrial development was concentrated in urban areas, which meant that the undisturbed development of ecosystems was possible outside these centres; for example at the border or in peripheral areas which were unsuitable for industrial production. In these areas the natural landscape could develop nearly without human impact and the biodiversity of plants of animals was growing. The political changes after 1989 throughout Eastern Europe have had profound consequences for nature protection. The pollution in general is declining, especially water and air pollution. But the diversity of species and the preservation of strictly protected areas are increasingly endangered due to unsustainable exploitation, land-use changes, increasing traffic, river regulations and the growing consumption of land. The accession to the European Union in 2004 promotes the idea of ease of economic actions and access to markets, but at the same time this contributes to the serious threats for biodiversity. Additionally, tourism development in peripheral areas is one of the recently emerged threats for biodiversity.

In 1997 the national government developed a national biodiversity strategy due to the signing of the Convention on Biodiversity in 1994. The following activities were formulated to be taken:

- All biodiversity must be conserved – preferably in situ
- Induced loss of biodiversity must be compensated for to the highest possible extent
- Diversified landscape must be maintained in order to sustain the variety of life at all levels
- Biological resources must be used in a sustainable way
- Every one must share the responsibility for conservation and sustainable use of biodiversity.

In 2002 the council of the Slovak Republic passed a new edition of the act on Nature and Landscape Protection (the first was from 1994) that aims to be in line with the relevant EU legislation and at the same time reflects the international concepts and agreements in the area of nature conservation. The act especially considered the Birds Directive and Habitats Directive with the consequences of making practical steps towards the establishment of NATURA 2000 in Slovakia. The act defines the complex nature and landscape protection with five degrees of protection according to the natural value. As shown in Figure 4.2 many parts of the country are protected. Consequently, 22.8 per cent of the whole territory is declared as either national park (nine sites) or protected landscape area (14 sites).

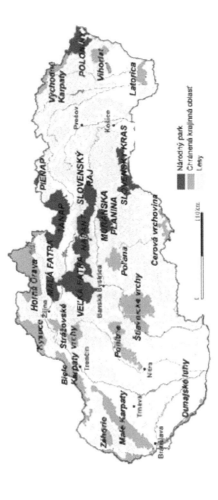

**Figure 4.2** Protected landscape areas (Chránená krajiná oblast), national parks (Národný park) and forests (lesy) in Slovakia

*Source: http://www.sazp.sk (2004)*

**Table 4.1    Biosphere reserves of Slovakia**

| Name | Area in ha | Year of dedication | National protection category |
|------|-----------|--------------------|-----------------------------|
| Slovensky Kras | 74,500 | 1977 | national park |
| Polana | 20,360 | 1990 | protected landscape area |
| Tatra | 105,650 | 1991 | national park |
| East-Carpathians | 40,778 | 1998 | national park |

Any remark on biosphere reserves is missing in the national law of nature protection. Therefore biosphere reserves are a kind of international label in addition to national protection categories. Nevertheless, in the middle of the 1970s Slovakia, formerly Czechoslovakia, founded a national MAB-committee and 1977 the first biosphere reserves were declared. The concept of biosphere reserves is different from other kinds of protection because biosphere reserves belong to the international network of the MAB programme by the UNESCO. In Slovakia, biosphere reserves are rarely seen as a tool for regional development. In contrast, on the international level this has been the case since the Seville conference from 1995. However, in Slovakia biosphere reserves have the primary purpose of conservation and nature protection with the exclusion of human activities. Consequently, the international status has little impact in the area and to relevant directives, laws and regulations. In other words, potential (legal) consequences from being biosphere reserve were overruled by the national protection category, for example national park and protected landscape area. Table 4.1 describes the four biosphere reserves existing in Slovakia.

All over Europe, a paradigm change of protected areas and nature protection has arisen. The new role combines nature conservation with the promotion of regional development. In contrast, in Slovakia this discussion might stay untouched. The traditional opinion that protected areas have as their main task, to preserve nature and to protect it from human impact, can be found commonly in the regions. Additionally, nature protection is often seen as inhibiting economic activities and the EU-legislation with for example environmental impact assessment as not supporting economic growth. On the other hand it should not be forgotten that with the rich biodiversity and a high natural potential, the landscape has a very good chance for a sustainable use for future generations. For regional development and especially for the local inhabitants it is not crucial what kind of label the protected area possesses but rather to what extent the implementation is successful.

**The Case of Slovensky Kras and Polana Biosphere Reserve**

In this section two out of four biosphere reserves in Slovakia are examined closely: Slovensky Kras and Polana. The following is mainly based on empirical studies of 2003. Two main methods were used: firstly, a standardized questionnaire for accommodation enterprises within the biosphere reserves and secondly, qualitative interviews with experts in tourism, regional development and local administration. Due to the small number of accommodation enterprises within both biosphere reserves and in their neighbourhood, the total number of questionnaires is relatively low (see Nolte 2005).

Firstly, the study areas were described in general and in their relationship to tourism, secondly the results of the survey follow with areas of conflicts in the biosphere reserves and finally, the potentials for regional development were analysed, for example the perception of the concept of biosphere reserves was examined and compared to the concept on international level.

## General Description of the Biosphere Reserves

*Biosphere Reserve Slovensky Kras*     The biosphere reserve Slovensky Kras is situated adjacent to the Hungarian Aggtelek biosphere reserve in the south east of Slovakia. Slovensky Kras is famous for its diverse karst phenomena. For instance, a series of plateaus ranging between 400 and 900 m above sea level are surrounded by steep slopes descending to adjacent basins, valleys and gorges. The region is famous, for instance because of the huge variety of dripstone caves, for example caves filled with ice and aragonite caves. Additionally, high biodiversity can be found, like species of butterflies and bats. Therefore, Slovensky Kras had already achieved the status of protected landscape area as early as 1973. Until 2002 this national status remained, when the efforts to establish a national park were successful. The boundaries enclose almost the same territory as the protected landscape area with about 35,000 ha. In 1977 Slovensky Kras was internationally declared as biosphere reserve, the first one in Slovakia. It covers an area of more than 75,000 ha, comprising landscape protected area and the national park. For the declaration a transition zone was defined.

Inside the biosphere reserve the population density is very low. The town Roznava (20,000 inhabitants) with cultural and administrative functions for the region, lies in the near vicinity. Roznava itself and also other towns in the region have a long tradition of mining and iron ore smelting. Today all mining activities have ceased but a lot of traces can be found in the landscape even now. The recent settlements and economic activities are concentrated in the basins and river valley outside the national park. In contrast to the adjacent Hungarian area of Aggtelek, the region has an industrial-rural character with industries exploiting and processing raw materials, machinery and heavy metal industry. Therefore economic structural weakness can also be stated as a typical characteristic of periphery.

*Tourism in the Biosphere Reserve Slovensky Kras*     Despite the good cultural and natural prerequisites for tourism activities, tourism has not played an important role in Slovensky Kras until today. Visitors have come to the area since the electrification of the Domica cave in 1932, which belongs to a greater cave system with extents to over 25 km underground. In 2002 the Domica cave had nearly 20,000 visitors. In addition, historic places are interesting for visitors, for example the baroque castle of Betliar with more than 50,000 visitors in 2002.[2] Summarizing, there are three main tourist activities in Slovensky Kras: 1. cave visiting, 2. visiting of cultural sites (castles) and 3. nature orientated activities like hiking and biking.

Unfortunately, reliable data about tourism development in Slovensky Kras is still missing as well as structural analysis of tourism. Available data suggests that tourism

---

2     According to data from the tourist information in Roznava.

has been growing since 1989. Short-term tourism is dominant in the region due to the vicinity of Slovakia's second largest city Kosice (240,000 inhabitants) which is the main source of visitors. Therefore incoming tourism is weak, only 10 per cent of the visitors come from abroad. The tourism depends strongly on the summer season (July and August). There are mainly private accommodation and smaller hotels. The regional centre Roznava offers several hotels and also the tourist information is located here. The tourist information works like a regional accommodation agency and in addition houses the local tourism board. In summary, the region is more of a local recreation and leisure place than an international tourism destination.

*Strengths* in tourism development in Slovensky Kras are the following:

- The cultural and natural attractions draw the attention of tourists
- The supply of beds is sufficient in Slovensky Kras for the recent tourists numbers
- Positive information about the region is provided by the tourist information centre in Roznava.

However, tourism in Slovensky Kras has *weaknesses*, too:

- Only a little money was invested in tourism with the consequence that most of the beds are of a low-standard
- Little information about the biosphere reserve is being disseminated.

*Biosphere Reserve Polana*    The biosphere reserve Polana is located in the central part of Slovakia in the proximity of the district city Banska Bystrica. The landscape was shaped mainly by volcanic processes more than 10 million years ago. The caldera with a diameter of up to 6 km is well visible, with altitudes of 1,300 m and up to 1,580 m (highest peak *Zadna Polana*). Additionally, the traditional agriculture formed a typical landscape with very small fields and the characteristically terrace-like shape. The extensive use of the landscape assures the high biodiversity in the area and consequently an area of 20,360 ha became protected landscape area in 1981. The international declaration of the entire area as biosphere reserve occurred in 1990 with a core zone of 1,338 ha, a buffer zone of 9,183 ha and a transition zone of 9,839 ha.

Within the biosphere reserve there are only three villages and few settlements with a total of 400 inhabitants. Most inhabitants are retired and those in employment mainly commute to industrial enterprises outside the biosphere reserve. There are only a few jobs in forestry and agriculture. In the south, on the verge of Polana, the town Detva has over 12,000 inhabitants and the central functions for the region. However, the largest enterprise (heavy industry) in Detva closed down in 2002 and more than a thousand workers lost their jobs. Problems within the biosphere reserve result mainly from the demographic situation, as younger people move out of the region and the remaining population has a high proportion of elderly people. The landscape is mainly characterized by a traditional, subsistence agriculture. Consequently, this area shows typical characteristics of structural weakness.

*Tourism in the biosphere reserve Polana*   Tourism is very weak in the Polana biosphere reserve. One hotel ('Polana Hotel') with 200 beds is centrally situated in the buffer zone of the biosphere reserve. The hotel offers three ski-lifts and other facilities like tennis courts, swimming pools and bowling. In the buffer and transition zone there are three localities with a total of 50 cottages and 20 recreation chalets of a poor standard. Altogether they have a capacity for 450 people. The landscape is very attractive for outdoor activities like hiking, biking, climbing, paragliding and winter sports due to 120 km of marked trails through the biosphere reserve. There are locations which are famous with climbers, with up to 10 m high rock formations. It is mainly an area for recreational use, only a small proportion of guests stays longer than one day, mainly at the 'Polana Hotel'. Because of the fact that there is no regular tourism survey, the total number of visitors can only be estimated. The biosphere reserve is visited by approximately 13,000 national and 5,000 foreign visitors each year (see Sorokova and Pichler 2002).

The *strengths* of the tourism development in Polana consist mainly of the following:

• Tourists favoured nature orientated activities for their stay
• The small number of tourists does not cause much damage to the natural environment
• Marked trails and three information points give tourists orientation in the biosphere reserve.

However, the following main *problems* can be listed:

• Especially bikers cause damage to vegetation because they illegally enter the areas with the highest degree of protection
• The special kind of soil (andosol) is very vulnerable to erosion. The soil lacks inner cohesion and can suddenly behave like liquid when it contains a high amount of water and is subject to physical impact
• The placement of the main tourist centre, the Polana Mountain hotel, is inconvenient due to being situated in the buffer zone and this causes pressure on the natural environment, e.g. traffic on the road.

*Sources of Conflict in the Biosphere Reserves*

In this and the following section, main results of the empirical studies provide information on sources of conflicts and potential for regional development by canvassing opinions of the local stakeholders.

*Guests' Interests*   The questionnaire of accommodation enterprises in both biosphere reserves gave more insight into the guests' interests in the opinions of their hosts, for example interests in different environmental and regional issues. Figure 4.3 illustrates the general low interest in typical environmental issues, for example using public transport, interest in environmental problems of the region and consuming ecologically produced food. Additionally, differences between the biosphere reserves

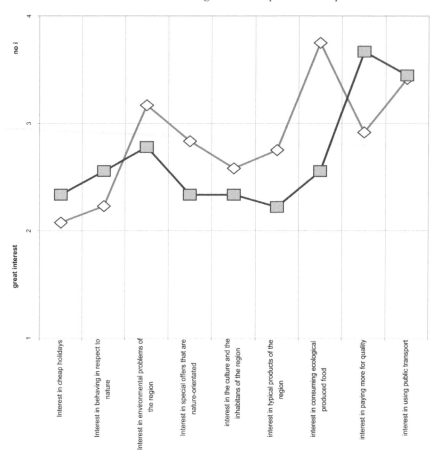

**Figure 4.3     Interests of guests according to statements of the accommodation enterprises**

*Source: Drafted by author*

were visible, for example in consuming ecologically produced food. The tourists greatest interest is in cheap holidays, in the opinion of the accommodation enterprises. Furthermore, having interest in regional products is typical for the tourists. One of the lowest interests was paying more for higher quality. As previously described, the tourism standard is at a low level and many accommodation enterprises can be described as simple. To conclude, the enterprises did not see a chance for higher priced tourist offers because of the low interest by their guests.

*Conflicts between Tourism Development and Nature Conservation*   It is very interesting to look at the results in regard to the conflicts between nature protection and tourism development (see Figure 4.4). The overlap of different interests is very likely. The answer to the question 'how would you rate the potential conflicts between nature conservation and tourism development?' was moderately in general; however, they differed between the regions. In Polana the respondents always estimated fewer

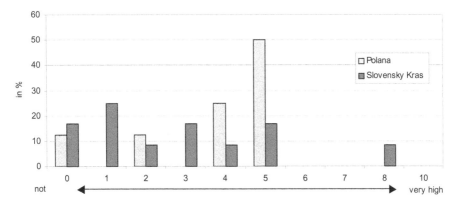

**Figure 4.4   Rating of potential conflicts between nature conservation and tourism development**

*Source: Drafted by author*

than six (on the scale from one: nonexistent, to ten: very high) and the majority valued the conflicts in the middle. To interpret, conflicts between tourism and nature conservation were seen, but due to the general low tourism intensity, they do not cause any pressure to react on. Further questioning confirmed this. Regional experts rated the current problems caused by tourism development as not very extreme. The emphasis was more on future threats. For instance, poor local interest in agricultural land could change the landscape that is an important tourism factor for Polana. Additionally, the lack of a proper sewage system in the villages could cause negative direct impact to the environment, if tourist numbers increase.

In contrast, in Slovensky Kras the answers revealed a greater variety; however the rating was generally lower than six. In Slovensky Kras there were more people rating the conflicts lower than in Polana, despite nine per cent of the answers suggesting that conflicts between nature conservation and tourism development in Slovensky Kras were high (rated as eight). Regional experts mostly agreed on the opinion that presently tourism did not harm the environment very much, but will possibly do so in future. Therefore, the tourism growth was seen sceptically by experts in nature conservation.

## Potentials of Regional Development in the Biosphere Reserves

### Role of Protected Area for Regional Development

The regional development in protected areas is limited by several regulations with protection purposes and these restrictions often result in severe conflicts. As previously mentioned, conflicts are weak in the biosphere reserves. According to statements of regional experts in both biosphere reserves, the relatively low conflicts are advantages to promote regional development by tourism development. Tourism is the only viable sector with the potential for growth among all sectors (including agriculture or industry). Especially in Slovensky Kras, the promotion of tourism development can rely on the long tradition of tourists' arrivals.

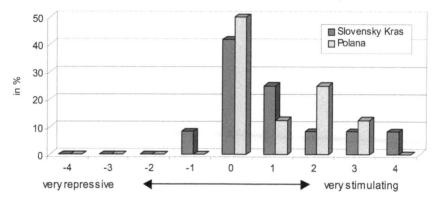

**Figure 4.5    Influence of the protected area on tourism development**
*Source: Drafted by author*

Until today, the protected area does not influence the tourism development very much. During the interviews, tourism enterprises commented on the influence of the protected area on tourism development (see Figure 4.5). Nearly 50 per cent thought that the protected area has no influence at all and rated the influence at '0' (on the scale from '–4' to '+4'). Only few answers were in the negative range whereas many thought that the protected area stimulates tourism. Furthermore, over half of the accommodation enterprises mentioned the protection of the area in their advertisements (in Slovensky Kras 61 per cent, Polana 55 per cent). Over 80 per cent of the accommodations additionally hand out information brochures about nature conservation, provided by the administration of the protected area. Taken together, the accommodation enterprises were convinced that a protected area could have a positive effect on tourism. However, the label biosphere reserve was not mentioned in any cases.

In contrast, the opinions of the administrations were not very positive concerning further tourism development. Many thought that tourists harm nature, directly or indirectly. They want to stop tourism development but cannot deny that tourism grew in and around the biosphere reserve. Visitor management is one of their tasks which they do by e.g. marking trails, installing information boards and producing information material. The aim is to guide tourists around the area whilst protecting the most vulnerable parts.

The situation differs between Slovensky Kras and Polana. The first has the advantage of being recognized as a national park, thus having more resources and employees. Polana has 'only' the status of protected landscape area and consequently its administration is not only responsible for this area. In addition, the administration office is located 20 km outside the area in Zvolen. In both cases, new strategies of how to cope with tourism development in a positive way and to find integrated approaches, are missing.

*Perception and Communication of the Biosphere Reserve Concept*

The concept of biosphere reserves promotes the idea of sustainable development. To reach this goal, it is crucial to communicate these aims within the region itself.

Therefore people working in tourist accommodation were asked in the questionnaire about their associations in connection with biosphere reserves. The answers are strongly connected to nature conservation or the natural landscape ('clean air', 'protection of plants and animals' and so on). Terms that express the integrated concept of biosphere reserve were seldom found. Additionally, it was obvious that the respondents did not make a distinction between the biosphere reserve and the national park (in Slovensky Kras) or the protected landscape area (in Polana).

Further investigations revealed that its international approval as a biosphere reserve was not communicated sufficiently within the areas. In Slovensky Kras even the young national park is better known by the public. Consequently, more people can associate something with a national park, while the term biosphere reserve remains mysterious. Good evidence for this is provided by the comparison with the biosphere reserve Polana, which is not approved as a national park. The situation here is dominated by the relatively low national status of protected landscape area which overlaps the international protection status. According to the statements of local tourism experts, few people know that the region is a biosphere reserve. As a matter of fact, some local inhabitants do not even know that Polana is a protected area.

In addition, for tourists spending their time in the area, it is hard to find any signs or information about the biosphere reserve. In Polana one sign post marks the area as a biosphere reserve; in Slovensky Kras any such signs are missing. However, in Roznava the protected area administration of Slovensky Kras held an exhibition about the biosphere reserve; unfortunately the opening hours are not very convenient. Furthermore, maps rarely indicate the correct size and location of the biosphere reserve Slovensky Kras. As a result, the transition zone has no function and hardly anybody knows about the transition zone.

Concluding, the attempts to inform people about the biosphere reserve are very weak in the regions and need to be improved.

*Decision Making Process and Role of Local Inhabitants*

Local inhabitants play a crucial role in protected areas. On the one hand, the acceptance of the protected area is mostly low by local inhabitants living in or in the vicinity of the biosphere reserve. This is caused by the fear of too many restrictions linked with nature protection. On the other hand, if they see benefit from the existence of the biosphere reserve, they will support nature conservation and the idea of sustainable development. Tourism can provide opportunities to benefit from the natural landscape. The questionnaire analysed whether the accommodation enterprises like to participate in decision making processes (see Figure 4.6). The overall readiness to participate is generally high. However, the differences between the two biosphere reserves are obvious. The broad agreement to this question was remarkable in the Slovensky Kras with approx. three quarters of the total responses being positive. Whereas in Polana slightly more than half of the respondents agreed. In Polana one third are uncertain about their participation in decision making. This may be the result of past experiences with participation procedures with some of them negative. An owner of a small hotel described the general scepticism as 'the talking didn't bring solutions'. However, a categorical refusal was stated only rarely.

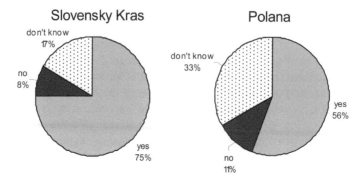

**Figure 4.6    Would you take part in decision making?**
*Source: Drafted by author*

Thus, there is an interest in taking part in regional decision-making, but not without conditions.

In contrast to this, many experts share the opinion that local people show little interest in regional development. According to statements in both biosphere reserves, it is very difficult to motivate local inhabitants to take part in processes which go beyond individual short-term profit. Therefore, the largest problem concerning regional development in biosphere reserves is the lack of initiatives from local inhabitants, who should be the main stakeholder for implementing sustainability. The reason for this situation is often identified as the previous socialist system: there might still be a strong faith in the capability of the state to regulate and finance at the regional level, as was usual in the socialist system.

In the region there is a divided group of inhabitants. On the one hand there are active and enthusiastic people who wish to develop the region with the help of tourism. These people are mainly residents who have not been living in the area for generations. On the other hand there are the main inhabitants of the villages situated close to the area who are not open minded and who are often suspicious of any changes in their life. Most of them have no interest in participating in decision-making processes because they do not see the connection between nature conservation and its value for the economic growth in the region.

It is a hard task to reach the local inhabitants, but in the long run there is no other chance for a sustainable future than to integrate them. This is not specific to these regions but can also be found in other rural areas throughout Europe.

For regional development the cross-linking between local stakeholders is of great importance.

It is particularly important in tourism if the goal is sustainable regional development. In both examined biosphere reserves there is a lack of networks, which is expressed in the experts' demand for the establishment of new networks and the maintenance of existing ones. If you have cross border cooperation, the opportunities to request financing in the European Union will be much more promising. Therefore, we can find international projects in the biosphere reserves financed, for example, by the European Union (EU) and the United Nation Environmental Programme (UNEP). Especially because of the common border with the Hungarian biosphere reserve Aggtelek, in Slovensky Kras bi-national as well as multi-national projects take place. Some of them are concerned with aspects of regional development and tourism with questions of how to improve cooperation.

In Polana, an active network is the Village Association of the so called 'Microregion Severne Podpolanie' with the aim of promoting tourism activities to increase the quality of life and employment in the region. Within the framework of Local Agenda 21, this network was started on the initiative of mayors of the villages and civic initiatives in the region of the biosphere reserve.

## Conclusion

Summarizing, the survey results suggested that the tourist enterprises do not see any differences between sustainable tourism and low tourism intensity. Severe conflicts are only seen in future if tourists increasingly travel to the areas. As shown in the case studies, the situation in biosphere reserves in Slovakia is dominated by conservation activities. Tourism is typically an untapped potential for regional development but will grow constantly, anyhow. Therefore, the future task is to focus on how to manage and control visitor flows and to take the opportunity for environmental education. For this purpose, biosphere reserves are perfect places to learn more about nature, especially for children.

The investigations indicated the important role of the administration of protected areas. Among others, their task is to manage tourists inside the biosphere reserve and to provide opportunities for environmental information and education. However, the examined administrations in the biosphere reserves had different resources, for example Polana has fewer employees and less financial resources. Therefore, Slovensky Kras is more capable of coping with the tasks of sustainability, but they have little power to implement ideas in the territory because in many cases the state is not the owner of the land. This is a serious problem for the administrations. In addition, even the major tourist attractions in Slovensky Kras, the dripstone caves, are managed by a national authority. As a result the administration of the biosphere reserve concentrates its activities more on environmental issues. To manage the region in a sustainable way the networking of all relevant organizations is urgently needed. In this context the administration could act as an umbrella organization to consider all interests and to harmonize them to benefit the whole region. However, the opportunity that lies within the international concept of biosphere reserves has not been recognized in the regions. Thus the thinking has to be changed urgently for administrations as well as for accommodation enterprises.

To analyse whether tourism promotes regional development, the case studies provided the perspective of the local inhabitants and regional stakeholder. Although tourism standards and quality are on a low level, many experts predict a constant growth in tourism in both regions. Even if many believe that tourism alone could not solve all regional problems, tourism remains the main hope for regional development. There is still potential to improve the link between tourism and nature conservation as a kind of ecotourism or even sustainable tourism. However, tourism experts in both regions do not distinguish between small-scale tourism and sustainable tourism.

Tourism is one of the key factors for sustainable development in the biosphere reserves examined, since this sector can be the economic engine for the conservation of cultural and natural values.

Tourism can also contribute to sustainable development in other ways: people are probably more likely to respect their own natural and cultural surroundings if they experience the value of these through being confronted by visitors looking for just that. This might also help to avoid the consumptive tourism developments – those that exceed the carrying capacity of the vulnerable natural and cultural landscape – like large scale ski facilities or extreme sports.

Until now, the biosphere reserve is not seen as a promoter for regional development itself. Biosphere reserves are seen more in the context of nature protection. The opportunity to implement the idea of sustainable development, as it is included in the Seville strategy for biosphere reserves, is not used effectively in Polana or in Slovensky Kras. For this purpose integrated approaches and their implementation are urgently needed.

## References

Altrock, U., Güntner, S., Huning, S. and Peters, D. (ed.) (2005), *Zwischen Anpassung und Neuerfindung. Raumplanung und Stadtentwicklung in den Staaten der EU-Osterweiterung.* (Berlin).

Baker, S. (2002), 'Environmental politics and transition', in Carter and Turnock (eds): *Environmental problems in East Central Europe*, 2nd edition. (Routledge, London), 40–56.

BMU, Bundesministerium für Umwelt, Naturschutz und Reaktorsicherheit (ed.) (2003), *Biodiversity and Tourism. The Case for the Sustainable Use of the Natural and Cultural Heritage of Banska Stiavnica, Slovakia* (Bonn).

BMWA, Bundesministerium für Wirtschaft und Arbeit (ed.) (2003), 'Tourismuspolitischer Bericht der Bundesregierung', *BMWA-Dokumentation* 521, (Berlin).

Bucek, M. (2000), 'Depressed regions – or depressed regional policy?', *Informationen zur Raumentwicklung* 7/8, 403–409.

Drgona, V. and Turnock, D. (2002) 'Slovakia', in Carter, F.W. and Turnock, D. (eds): *Environmental problems in East Central Europe*, 2nd edition. (Routledge, London), 207–227.

European Commission (ed.) (2003), *Tourism – Europe, central European countries, Mediterranean countries. Key figures 2001–2002.* (eurostat, Luxembourg).

Grotz, F. (2000), 'Politische Institutionen und post-sozialistische Parteiensysteme in Ostmitteleuropa. Polen, Ungarn, Tschechien und die Slowakei im Vergleich', *Junge Demokratien* 5. (Leske+Budrich, Opladen).

Mariot, P. (2003), 'Nouvelles tendances du développement du tourisme et modèle régional de la republique Slovaque', in *Hommes et Terres du Nord*, 2003/4, 18–27.

MHSR, Wirtschaftsministerium Slowakei (o.J.) (undated), *Slovakia Tourism Strategy. Progress report.* (Bratislava).

Netzhammer, M. (2005), 'Zäsur im Paradies', in *akzente* 1.05, 40–43.

N.I.T., Institut für Tourismus- und Bäderforschung in Nordeuropa (2004), *A new Tourism for a new Europe – Ein neuer Tourismus für ein neues Europa. Startdokumentation zur Fachkonferenz auf der Internationalen Tourismusbörse Berlin.* 13. März 2004. (Kiel).

Nolte, B. (2005), *Tourismus in Biosphärenreservaten Ostmitteleuropas. Hoffnungen, Hindernisse und Handlungsspielräume bei der Umsetzung von Nachhaltigkeit.* Dissertation. (Mensch&Buch Verlag, Berlin).

OECD – Organization for Economic Co-operation and Development (ed.) (1996), *Regional problems and policies in the Czech Republic and the Slovak Republic.* (OECD, Paris).

Robertson-Wensauer, C.Y. (ed.) (1999), 'Slowakei: Gesellschaft im Aufbruch', (Nomos Verlag, Baden-Baden).

Ságát, M. (1998), *Polana nad Detvou.* (Knizne Centrum, Zilina).

Slowakische Zentrale für Tourismus (2004), *Pressebericht anlässlich der Internationalen Tourismusbörse Berlin,* 12.–16.03.2004. Deutsche Vertretung in Berlin.

Sorokova, M. and Pichler, V. (2002), *Case study. The tourism potentials and impacts in protected mountain areas. Polana protected landscape area – biosphere reserve, Slovakia.* (Centre for Scientific Tourism in Slovakia, Zvolen).

Spisiak, P. (2003), 'Transformation des régions rurales, privatisation, changement de la production agricole en Slovaquie', in *Hommes et Terres du Nord,* 2003/4, 42–65.

Statistical Office of the Slovak Republic (ed.) (2001), *Population and Housing Census.* (Bratislava).

Tickle, A. and Vavrousek, J. (1998), 'Environmental politics in the former Czechoslovakia', in Tickle, A. and Welsh, I. (eds): *Environment and society in Eastern Europe,* 114–45. (Addison Wesley Longman, Harlow).

UNESCO (2002), *Biosphere reserves. Special places for people and nature.* (UNESCO, Paris).

Vollmer, M.M. (1997), *Ökotourismus – Ein Beitrag zur nachhaltigen Regionalentwicklung in der östlichen Slowakei am Beispiel des Nationalparks Slowakisches Paradies.* Unveröffentlichte Diplomarbeit an der Universität Trier, (Trier).

Williams, A.M. and Balaz, V.V. (2000a), 'Privatisation and the development of tourism in the Czech Republic and Slovakia: property rights, firm performance, and recombinant property', in *Environment and Planning A* 32, 715–34.

Williams, A.M. and Balaz, V. (2000b). *Tourism in Transition. Economic Change in Central Europe.* (Tauris Publisher, London, New York).

Chapter 5

# Nature Parks and Regional Development in Austria: A Case Study of the Nature Park Ötscher-Tormäuer

Christine Gamper, Martin Heintel, Michael Leitner
and Norbert Weixlbaumer

## Introduction: The European Context

Since the 1970s nature parks in Europe have enjoyed an uninterrupted boom (see FNNPE 1994, 6). Nature parks in the wider sense are those protected areas that can be classified as IUCN-Category V. They comprise approximately one half of all protected areas in Europe in 2006 (see Chape et al. 2003, compare Figure 1.1). However, the range of all protected areas classified as Category V in Europe is large. At the national level (Austria) the protected areas falling under this category include national parks and regional parks, regional nature parks and nature parks as well as protected landscapes (see Weixlbaumer 2005a). Category V is defined as a protected area that 'has as its main management objectives the protection of the landscape or the marine area, in addition to serving the people's recreation'. It is an area that, 'throughout time and through the interaction of humans with nature, has formed a landscape with a unique character, outstanding esthetic, ecological and/or cultural values and with an often remarkable biological diversity. The uninterrupted continuation of this traditional interaction is imperative for the protection, preservation and further development of this region' (see Europarc and IUCN 1999, 30).

Nature parks in Europe have slightly different names and meanings. For example, in Austria a nature park is labeled as a 'landscape' with its legal status based on the decree of a (mostly) landscape protection or nature conservation area. It is a protected landscape area, which has developed through the interaction of man with nature and which has no or little wilderness. In Austria, nature parks do not belong to any official zoning. The management structure is formed by the responsible state government (Department of Nature Conservation) or by a regional management in cooperation with a local nature park association.

What are the reasons for the continued attractiveness of nature parks? Nature parks are, according to the above definition, instruments that can combine the protection and the use of a landscape (see Hammer 2003, Mose and Weixlbaumer 2003). With the wake up of the environmental awareness since the 1970s and the increasing demand for 'leisure areas', the nature park concept gained acceptance and subsequently adjusted to various developments. Today, nature parks, due to their integrative approach, conceptually fit the dynamic-innovative nature conservation paradigm (compare chapter 1).

The following discusses the nature park concept and its potential for regional development using Austria as an example. In addition, a nature park in the Austrian province of Lower Austria, the Nature Park Region Ötscher-Tormäuer, will be presented as a case study. At the end of this chapter, strategies for an integrative regional management based on an analysis of this case study will be discussed.

## Nature Parks in Austria: Historical and Paradigmatic-Functional Development

Nature parks in Austria, as a small Central-European country, are characterized by their small size. Sizes range from 17 to 58,000 ha. From a historical perspective, the development of nature parks in Austria is similar to other European countries, including Italy, France or Germany. In 1962 the first Austrian nature park, Sparbach bei Mödling (approximately 15 km south of Vienna) was founded. Recreation close to Vienna and (day) tourism were, during the founding days of nature parks, the important objectives. The first nature parks were small and close to urban centers. They represented, as for example the landscape with pine forest and lime stone cliffs, located south of Vienna, cultural and natural landscape types. From the beginning nature parks were not representatives of the classic-static paradigm, although conservation was in some parks the main objective: this was also true for the Nature Park Ötscher-Tormäuer, when it was founded in 1970.

In 2006, Austria possessed 40 nature parks, which together cover an area of 350,000 ha and about 4 per cent of the size of the national territory. According to the Association of the Austrian Nature Parks, they can be described with the following features (Handler 2000, 43): 'Broad acceptance among the population; not only differences in size, but also in staff, equipment, and finances; voluntary contribution of many employees and mostly located in eastern Austria.' (22 nature parks alone are located in Lower Austria; see Figure 5.1.) Nature parks are basically free of charge and entrance fees are on a voluntary basis or for special attractions (for example game parks).

During this founding period, nature parks were mostly recreational regions for the Viennese urban agglomeration. The recreational function 'being in nature' was combined with a rather modest concept of nature conservation and landscape protection. This is evident from the first nature parks in Lower Austria (see ÖGNU and Wolkinger 1996). Within the scope of their recreational objective, nature parks pursued – although rather modest at the beginning – nature- and environmental education for a broad audience. Nature parks developed into 'educational landscapes' with the establishment of educational trails and information boards. At the beginning, the pedagogical implementation of this environmental education lacked value added landscapes – a few small nature parks insufficiently represented the Austrian types of landscape. On the other hand, the implementation lacked political will power, which

# Nature Parks in Austria

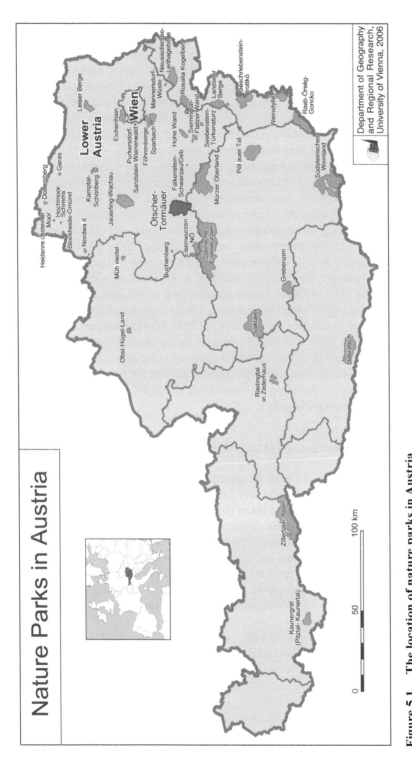

**Figure 5.1 The location of nature parks in Austria**
*Source: Drafted by authors after Zollner and Jungmeier 2003. 16*

became evident with marginal financing and care. In addition, the local population was rarely included into the politics of nature parks. Altogether, awareness for an integrative nature conservation and landscape protection was missing.

Not until the period of the institutionalization of nature parks through the constitution of an Association of the Austrian Nature Parks in 1995 and an international revaluation of this protected area category an intensification and coordination of activities has been taken place. The coexistence of the three nature park functions, nature conservation, recreation, and education, which complemented each other and was strongly believed in, was soon extended. Nature park areas should contribute to the regional development of rural areas. In accordance with a further paradigmatic development, a coexistence of the functions nature conservation, recreation, education, and regional development was subsequently postulated. Today, nature parks have become a factor for integrative regional development. In this context, the Austrian State of Styria initially demonstrated innovativeness. It not only established relative systematically extensive nature parks, but also equipped these nature parks with, by Austrian standards, solid infrastructure, including a base budget, a (municipality independent) manager, a nature park academy, a public relations office, and so on. In contrast, the State of Upper Austria focused on a nature park development that comes from the region itself (for example, a bottom-up development; the Nature Park Mühlviertel). Therefore, the province of Upper Austria possesses few, but well-functioning nature parks with model character. In the future, nature parks in Upper Austria will cover all large cultural landscape types in the province. In addition, the State of Tyrol with the concept of 'protected areas that are being managed' shows innovativeness today. Therefore, an increasingly systematic process is recognizable in the Austrian nature park politics.

**Nature Parks as Instruments of Regional Development**

Nature parks can lay claim to be described as model regions for sustainable development. On the basis of the four functions (protection, recreation, education, regional development) a detailed list of objectives is drawn up (see Figure 5.2). With the additional function of regional development the Austrian nature park concept is in line with claims of other European nature park concepts. This has been an important step for the public's perception and discussions about nature parks in Austria.

However, nature parks can only become model regions in rural areas, if it is possible that, based on integrated sustainability, they can successfully become efficient regional planning instruments. This requires that:

- Based on the most recently discovered management goals, regional development, education and conservation should be treated equally and
- Nature parks should be positioned a new politically and in the awareness of decision makers.

For this reason, the International Commission for the Protection of the Alps (CIPRA) discussed at a conference titled 'Who is Frightened of Protected Areas? Protected

| Nature parks – equal coexistence of … | | | |
|---|---|---|---|
| **Protection** | **Recreation** | **Education** | **Regional development** |
| The objective is to protect nature through sustainable use of its diversity and beauty and to preserve the cultural landscape that has been formed throughout centuries. | The objective is to offer attractive and clean recreational facilities, which comply with the protected area and the character of the landscape. | The objective is to experience nature, culture, and their interactions through interactive forms of understanding and experiencing nature and through special offers. | The objective is to provide impulses for a regional development in order to increase the value added and to secure the quality of life for the population. |
| • visitor control | • hiking trails | • theme trails | • cooperation between conservation, agriculture, tourism, trade and culture |
| • Information about natural history | • cycling and bridle path | • adventure tours | |
| | | • information centers | |
| • 'soft' mobility | • resting places | • nature park school | • jobs through nature parks |
| • management of protected areas | • 'adventure' play grounds close to nature | • attractions specifically designed for target groups | • socially and environmentally compatible tourism |
| • research projects | • family friendliness and accessibility for the handicapped | • seminars, courses, exhibitions | • nature park products defined by specific criteria |
| • contract about environmental protection | • fresh air and peacefulness | • preservation of customs | • nature park restaurants |
| **Nature parks are model regions for sustainable development** | | | |

**Figure 5.2     Strategies of Austrian nature parks – four functions and goals**
*Source: VNÖ 2005, 4*

Areas as a Chance for the Region' (see CIPRA-Austria 2002) the current nature park policy. Politicians, regional managers, experts in the field, and individuals from the local population determined that nature parks and similar protected areas, including biosphere parks, are equivalent to the well-known protected area categories (first and foremost national parks) and therefore should be appropriately treated and financed. A crucial realization was that this fact has been advocated by the IUCN for a long time. This needs to be better communicated to the public, especially to politicians making funding decisions. Because only with appropriate idealistic and financial support, that is, the establishment of an appropriate management structure, nature parks and similar protected areas under the IUCN Category V can meet their presently high standards and become in reality model regions for sustainable development.

Which basic prerequisites need then to be in place for nature parks to provide impulses for rural development or even serve as model landscapes (see Weixlbaumer 2005b)?

The deciding factors for an exemplary nature park development are human capital (including the knowledge of the local population) and the landscape. Both factors need to be appropriately activated, that is, their values need to be taken advantage of. Because nowadays nature parks are expected to offer their visitors an entire spectrum of attractions (compare the discussion in Gill 2003, as referenced in Reusswig 2004, 149). This includes a harmony aspiring human-nature-relationship, which is manifested, for example, in the search for peacefulness and balance. It also includes the effort to satisfy 'the hunger for adventure' in nature, for example, to experience the thrill in the mountains.

These basic factors for a nature park adequate (that is, integrative) regional development can only be effective with appropriate political will. Political will can initially be expressed through the execution of existing (nature conservation) laws. Political will power also means taking advantage of the existing network of interest groups, including NGO's and private business people. Overall, it is the goal to regulate and implement an entire package of fundamental ideas. This needs to be put on a legal, planning, and financial basis. Built upon this is the functioning of the park that has a sufficiently large size, zoning laws, if necessary, and sufficient personal. A general rule for Alpine parks, dependent on the type of landscape, is one person per 2,000 ha (see Scherl et al. 1989). As such, a nature park can become an efficient business for a region. In addition, a number of voluntary and part-time employees are needed. The number of employees is dependent on the seasonally adjusted requirements of the park.

Based on numerous studies (compare, among others, Heintel 2005, Weixlbaumer 2005a, Mose and Weixlbaumer 2002) during the last ten years in the Department of Geography and Regional Research at the University of Vienna, it has been shown that the relationship between area management, in the context of regional protection, and the associated measures of regional development, has increasingly become more important. However, the field of applications for a regional management with direct connections to strategies for regional protection has been paid insufficient attention overall. Newly planned or recently established national parks (that is, protected areas with the highest IUCN category of protection) are increasingly incorporated into regional development concepts (for example, National Park Kalkalpen, National Park Gesäuse).

## Case Study Nature Park Ötscher-Tormäuer

The nature park was established during the beginnings of nature park policies in Austria and during the wake of the European Year of Nature Conservation in 1970. Based on a landscape protection area it served as an instrument to prevent the building of a hydroelectric power plant. As an economically value added instrument of the agro-tourism region Ötscherland, it was never really effective. The park is located within the (daily) catchment's area of three provincial capitals, including Linz (Upper Austria), St. Pölten (Lower Austria), and Vienna (Vienna). This area is an agglomeration with a total population of approximately three million. The central massive of the park region is the 1,893 m high Ötscher, which is a far visible and identifiable landscape feature. The nature park is inhabited (approximately 800 people live inside the park) with an important landscape potential, mixed ownership and a consensus-seeking landscape protection. During a process of enlargement in 2001 the area of the nature park doubled and increased to 14,869 ha. After that an additional enlargement to approximately 17,000 ha followed that included parts of the municipality Lackenhof, west of the Ötscher summit.

At the beginning of 2006, both enlargements have been informally executed between the state (province) and the municipalities, but have not been embodied into state law. The enlargement of the Nature Park Ötscher-Tormäuer should thus have come to an end. Future visitors can therefore increasingly experience the nature park as a 'wilderness nature park' that includes a series of tourist attractions (for example, a stalactite cave, water falls, gorges, pure watercourses, forests and alpine pastures, brown bears, and so on).

As in most Austrian nature parks, the legal authority is an association that includes the mayor from each municipality, located within the park. Until 2003, the staff and financial support was low. Therefore, the park portrays both the problematic and hopeful image of Austrian nature parks. With the implementation of the investment program (see ETB 2002) and the temporary employment of a project supervisor (until the beginning of 2006), the possibilities of a nature park management changed positively over a period of three years. Starting in summer 2006, the support for a professional management and marketing is again uncertain.

How is it still possible to utilize natural rural areas for regional management with especially protected areas becoming development factors (development engines) in underdeveloped regions? Our study area is the Nature Park Ötscher-Tormäuer (see Figure 5.3), which is situated in the Eisenwurzen, a region that stretches across several state boundaries (along the Lower Austrian, Styrian, and Upper Austrian Eisenstrasse). The nature park is distinguished not only by comprehensive regional funds from the EU (the funding period is from 2000–2006: Objective 2 Region or transition region from Objective 2 or 5b-old; LEADER+ regions), but also by numerous adjacent protected areas. The Nature Parks Lower Austrian and Styrian Eisenwurzen, together with the landscape protection area Dürrenstein-Ötscher, the National Park Kalkalpen and the National Park Gesäuse founded in 2002 form together an extensive system of protected areas. This entire region lies adjacent to the project Eisenstrasse, which stretches across provincial boundaries (see Heintel and Weixlbaumer 1998). Although this region has shown massive decline since World War II, it has experienced strong economic impulses through measures of regional development more recently. The Nature Park Ötscher-Tormäuer is the oldest established protected area in this region nevertheless it still lives in the shadows.

In this case regional management must not only develop suitable concepts for area management, but also coordinated measures for an integrative strategy of a set of measures for the economy, tourism, and marketing for the entire region. The southern enlargement of the nature park in 2001 (includes the region of the Ötschergraben) can already be interpreted as a regional effort of the affected municipalities to attach greater importance to regional protection in connection with measures of regional development in the future.

A regional management is challenged twice in this context. On the one hand, it needs to strengthen dynamic strategies for regional protection in those areas, where too little has been done previously to manage the functions protection, use, education, and regional development and to coordinate them with each other. This does not only address the conceptual and marketing-oriented area, but also the coordination

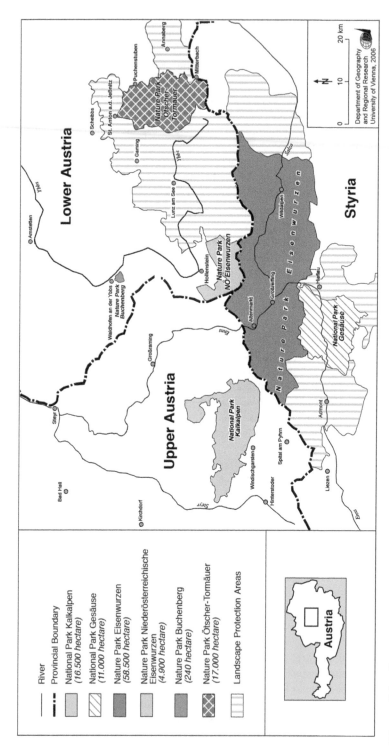

**Figure 5.3 Protected area projects crossing provincial boundaries of Lower Austria, Upper Austria and Styria**
*Source: Drafted by authors*

and networking between the municipalities that are located inside the park. On the other hand, this addresses an integrative-coordinating activity at the management level between the already mentioned adjacent protected areas that is a coordinated commitment of support from funds targeting specific regions and community initiatives. Additionally, all this needs to be carried out across provincial boundaries.

## Selected Results

How does the local population perceive the Nature Park Ötscher-Tormäuer as an engine for regional development vis-à-vis comprehensive development goals?

During a two-year project conducted by the Department of Geography and Regional Research at the University of Vienna two different studies titled 'The Nature Park Ötscher-Tormäuer in the Minds of the Local Population – Chances for Attempts of Regional Development' were conducted in five nature park municipalities, including Annaberg, Gaming, Mitterbach, Puchenstuben, and St. Anton an der Jeβnitz.

During the first study alone 807 interviews were conducted among the local population (approximately 12 per cent of the total population of the municipalities). The focus of these interviews included questions about the spatial perception and cognitive mapping of the boundaries of the nature park (using mental maps), qualitative perceptions towards the park (using perceptional profiles), and the knowledge and perception of how well the nature park fulfills the four objectives (protection, recreation, education, and regional development). Additionally, the interviewees stated their opinion about possible future development scenarios for their region (economy, tourism, and so on) and about different strategies for regional development (for example, nature park, LEADER+). Finally, the interviewees took the opportunity to express their wishes about the nature park and its management (using open-ended and structured questions).

The results from the first investigation showed that the local population perceived regional development and education as less important goals (as compared to protection and recreation) for the region Ötscher-Tormäuer. For this reason, the second investigation studied the goals of regional development and education in more detail. Therefore, the focus was on the 'marketing of the region' and the nature park's role in this context ('Which landscape features and regional images are created, communicated and which are effective?'). A second subject area was 'role model and conflicts of utilization' based on an analysis of different role models of the region (role model of tourism development of the Nature Park Ötscher-Tormäuer, role model of the development Eisenstraβe, Ötscherland, Mostviertel, and so on) and the representation of conflicts of utilization in the politics of regional protection, in general and based on the functional diversity of the nature park, in particular. On the other hand, questions about the contribution the (nature park) landscape provides for 'education' (strengthening of the educational mission and drafts for a strategy for educational measures in nature parks) and the nature park regions serving as model regions for a sustainable development in rural areas (experiences from other nature parks, for example, as part of a regional competition about cultural landscape and possibilities for the Ötscher-Tormäuer region) were raised.

The goal of the following discussion is not to present this study in detail, but rather to focus on possibilities and limitations regional management faces with respect to regional protection.

## Degree of Familiarity and Image

As a result, 98.6 per cent (796 out of 807) of the interviewed local population in all five municipalities said that they would know the Nature Park Ötscher-Tormäuer and that they would regularly frequent it with activities such as hiking, walking, cycling, hunting, fishing, skiing, and so on. The interviewees indicated that they practically live inside the nature park region. They identify themselves with the nature park, although specific knowledge about its planning are often missing. In addition, the 807 interviewees from the first study describe, on average, the Nature Park Ötscher-Tormäuer as 'quite familiar'. Above all, the good image of the nature park is expressed with the following perceived characteristics: very 'natural', 'invitational', 'attractive' and 'important' as well as very 'interesting', 'colorful', and 'lively'.

Of course, such a high degree of familiarity and positive image are a good basis for additional management- and planning measures in the Nature Park Ötscher-Tormäuer. Favorable conditions seem to be in place that in the future a financially and structurally better equipped nature park policies will pursue with determination the desirable functional diversity between protection, recreation, education, and regional development.

Cross-sectional topics of regional and regional planning importance about the nature park can be laid out, formulated, and implemented in detailed (for each municipality independent) projects. These are the exact tasks of a regional management. An active and coordinated engagement can result in an impulse and a revaluation of the region.

## The Nature Park: Associations and Expectations

The interviewees from the five municipalities located within the nature park associate the term 'nature park' foremost with characteristics, such as nature and (nature) conservation and less frequently with recreation, tourism, and landscape. Accordingly, the interviewees understand the central functions of nature parks mostly with broadly defined (nature) conservation, expressed in the protection of living space, regional protection and nature conservation. In contrast, all municipalities do not necessarily consider tourism, and especially education and regional development as characteristics to be associated with the nature park. This means that the regional population does not yet perceive an integration of the four functions (that is, conservation, recreation, education, and regional development), which has been demanded from the Association of the Austrian Nature Park's politics.

The two functions most often associated by the interviewees with the nature park, namely nature- and environmental conservation, are in reality associated more with nature conservation regions and national parks, rather than with nature parks. Only a well-organized regional management can satisfy this wishful thinking of the population.

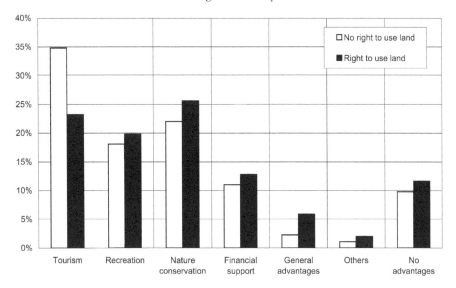

**Figure 5.4    Positive expectations by categories (in %) for all municipalities by 'right to use land'**

*Source: Drafted by authors*

The general expectations towards the nature park are with approximately 60 per cent positive. 20 per cent of the interviewees expect rather disadvantages, and 20 per cent do not have any expectations towards the nature park. Comparing across all individual categories, the expected advantages towards the nature park are predominant. Especially tourism, nature conservation, recreation, financial and economic funding are rated as the most positive categories (see Figure 5.4). However, the portion of negative expectations is with 20.3 per cent (281 out of 1,383) relatively high and includes conflicts of use, pollution (noise, waste, and so on) and too many tourists. This can mostly be explained by limitations (zoning, conservation regulations, conditions for cultivation, and so on) that regulations of the nature park, in the opinion of the interviewees, could involve.

When the results are differentiated by interviewees with or without having the right to use land in the nature park, then hardly any significant differences within the 'advantages' categories can be observed. However, answers differ when 'disadvantages' categories are considered (see Figure 5.5). People who do not have the right to use land in the nature park perceive negative consequences of the park less often (that is, they expect more often the category 'no disadvantages'), whereas people with the right to use land perceive more often 'conflicts of use' – predominantly in the form of restrictions for building projects and agriculture.

Since portions of the municipalities Annaberg and Mitterbach were in discussion to be part of the park enlargement, the detailed results for the 'disadvantages' categories are here especially interesting. Both municipalities record – with more than 25 per cent disadvantages each – a relatively skeptical position towards 'conflicts

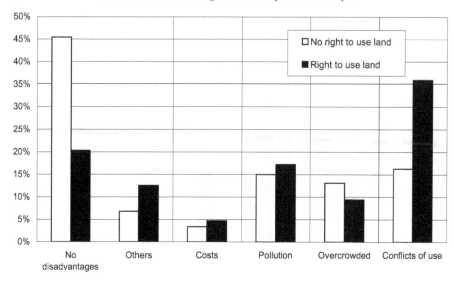

**Figure 5.5    Negative expectations by categories (in %) for all municipalities by 'right to use land'**

*Source: Drafted by authors*

of use'. Apparently, many uncertainties and fears, as well as information deficits regarding future regional management had existed here. In both municipalities, the fear of too many tourists and larger user groups is rated much higher compared to all other municipalities. Whereas this category averages only 12.8 per cent in all five municipalities, it is 25.3 per cent in Annaberg, 20.8 per cent in Mitterbach, but only 3.4 per cent in Gaming.

Positively perceived categories:

1. Tourism: hiking, tourists
2. Recreation: quiet countryside, enjoyment of nature, walking
3. Nature conservation: conservation of species, protection against urban sprawl
4. Financial support: regional financial support, monetary financial support, financial support for restaurants, economic advantages
5. General advantages: general advantages for the region, stimulation for the region
6. Others
7. No advantages.

Negatively perceived categories:

1. No disadvantages
2. Others

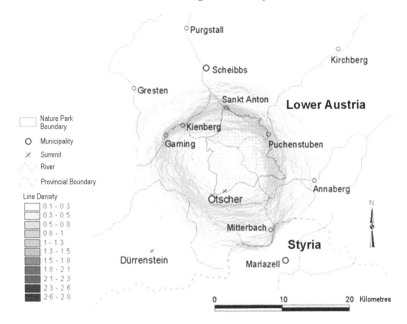

**Figure 5.6**     **The nature park as a 'giant' – the size of the Nature Park Ötscher-Tormäuer is perceived as being much larger than it is in reality by the local population (n=758)**

*Source: Drafted by authors*

3. Costs: costs for maintenance and infrastructure
4. Pollution: noise, waste
5. Overcrowded: too many tourists, mountain bikers
6. Conflicts of use: limitations for agriculture, landowner, development area.

*The Nature Park Ötscher-Tormäuer as a 'Giant'*

The evaluation of all 758 hand-drawn mental maps of the boundaries of the Nature Park Ötscher-Tormäuer shows an interesting result: In the perception of the population of the municipalities with an average of 23,286 ha the nature park is much larger than the actual size before the enlargement of the park in 2001. The actual size of the park is with 7,586 ha only one third of the perceived area. The interviewees thus overestimated the area of the park by almost twice its size (see Figure 5.6).

Jointly responsible for these results was the ongoing discussion about a southern enlargement of the nature park at the time of the study (including the municipalities Annaberg and Mitterbach). Many were unsure or assumed that the enlargement of the nature park had already taken place at the time of the study. The enlargement by the two municipalities was not carried out until after the interviews – however, smaller than its originally planned size. Only a smaller portion of the municipality Annaberg was integrated into the nature park. Independent of this, the regional

**Table 5.1    Achieving the objectives of the nature park, all municipalities (n=807)**

| Achieving the objective | Yes | | No | | No answer | | Total | |
|---|---|---|---|---|---|---|---|---|
| | Absolute | % | Absolute | % | Absolute | % | Absolute | % |
| Nature conservation | 747 | 92.6 | 30 | 3.7 | 30 | 3.7 | 807 | 100 |
| Recreation | 766 | 94.9 | 20 | 2.5 | 21 | 2.6 | 807 | 100 |
| Education | 396 | 49.1 | 310 | 38.4 | 101 | 12.5 | 807 | 100 |
| Regional development | 454 | 56.3 | 257 | 31.8 | 96 | 11.9 | 807 | 100 |

population before 2001 already considered the Ötschergräben – part of the 2001 enlargement – the heart of the nature park.

In addition, studying the line density provides insights into which map elements the interviewees used for orientation in drawing the nature park boundary.

Noticeable is the band of higher line density that extends from Puchenstuben via St. Anton an der Jeβnitz to Gaming. This is also where the misjudgment in the size of the nature park is apparent: whereas the actual boundary of the nature park runs from St. Anton an der Jeβnitz due south, the interviewees often drew the boundary to the west via Kienberg to Gaming.

South of Gaming until south of the Ötscher is a zone lower line density. Apparently, the interviewees have oriented themselves closer to the course of the Ötscher stream, rather than the Ötscher summit, which at the time of the interview (before the enlargement) was an element of the actual boundary.

Much higher densities for the cognitive southern boundary of the nature park were found along the Styrian-Lower Austrian provincial boundary southwest of Mitterbach. The reason for this is that, on the one hand, the provincial boundary is represented as a very prominent element in the map and, on the other hand, a nature park boundary along a provincial boundary, especially in federal Austria, is perceived as being very plausible.

Between Mitterbach and Puchenstuben a somewhat broader band of higher density can again be recognized, which is also very close to the actual boundary.

*Regional Development and Education*

Regional development is included as an explicit goal in the nature park concept. The question now remains to what extent has this goal been achieved compared to other more predominant goals?

Interviewees stated that nature parks should fulfill several functions, including environmental protection and protection of species, recreation and education, as well as stimulation of tourism and regional development. At first sight, these results cover the entire spectrum of all functions of nature parks as defined in the concept of the VNÖ (see Association of the Austrian Nature Parks 2005) (see Table 5.1).

However, a closer look at the results shows that in the opinion of the interviewees some functions are more important than others. This leads to an uneven distribution across the different categories attributed to be functions of nature parks: (nature) conservation (environmental protection and protection of species) and recreation

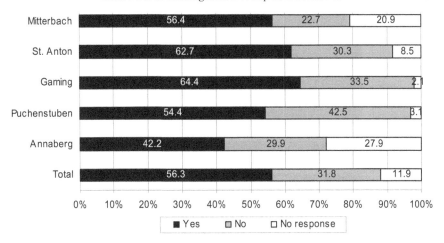

**Figure 5.7    Comparison between municipalities in achieving the goal 'regional development' (n=807)**

*Source: Drafted by authors*

were mentioned most frequently. Although education has always been a central theme in nature parks that has been visible in the form of information boards and educational trails, education was mentioned less frequently. Additionally, there is little awareness among the interviewees that nature parks play a substantial role in regional development and that regional development is becoming increasingly important.

When results are differentiated by highest completed degree of the interviewees then academics mention the categories education (11 per cent) and regional development (5.4 per cent) twice as often as compared to interviewees with less school education (5.7 per cent and 2.6 per cent respectively). In contrast, interviewees with completed mandatory education believe more often (12.5 per cent) than other interviewees (8 per cent) in tourism to increase the attractiveness of nature parks. While the nature park functions 'environmental protection' and 'recreation' are mentioned less often by academics compared to the other interviewees (difference of approximately 8 per cent), the 'protection of species' is mentioned more frequently. This may be due to a more differentiated knowledge and perspective of academics.

Overall, this analysis also shows clearly the different distribution in the individual task-categories and the level of information of the population (including all educational levels) across the functions of nature parks. A closer look at the factors 'regional development' and 'education', shows the following (see Figure 5.7).

More than half of the interviewees (56.3 per cent) perceive regional development tasks to be an objective of the nature park. Most likely this is due to the attractiveness for tourism, the use of the brand name and the possibility to receive financial support (EU-Structural Fond – Objective 2 Region, previously Objective 5b). The objective of the nature park to strengthen regional development is rated very high by the interviewees in Gaming (64.4 per cent) and in St. Anton an der Jeβnitz (62.6 per cent), but very low in Puchenstuben (42.5 per cent). Because of the difficult economic

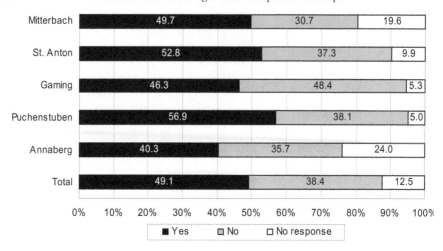

**Figure 5.8    Comparison between municipalities in achieving the goal 'education'**

*Source: Drafted by authors*

situation in the region, there are too few noticeable positive effects from regional development in the nature park (new jobs, increase in overnight stays, stabilization in the number of agricultural and catering businesses). The study concludes that there are also too few initiatives and activities for improving the regional value added of the nature park and for an integrative development in the region (marketing, tourism offers, co-operations, raising of people's awareness, and so on).

The interviewees – with the exception of Puchenstuben – rate the educational function (see Figure 5.8) of the nature park with the highest inefficiency. Especially in Gaming more interviewees (48.4 per cent) believe that the nature park has not yet achieved the 'education' objective. The satisfaction with this objective is also below average in Annaberg (40.3 per cent), with many interviewees (24.0 per cent) not providing an answer to this question. The discrepancy between the nature park's self-image and how the local population perceives the nature park is remarkably high. All activities offered by the nature park – guided tours, information boards, educational trails, theme trails, and so on – should be more attractive and complemented with additional creative and group-specific (children/adults, gender, specific interest) educational offers. In addition, special co-operations, for instance with schools (environmental education), hotels offering seminars, such as the 'Karthause Gaming' and the 'Alpine Hotel Gösing', or local businesses (agriculture) with their regional know-how could provide new and interesting avenues for the local population (raising of people's awareness, transfer of knowledge, continued education) as well as for tourists.

The local population from the region Ötscher-Tormäuer hardly recognizes regional development and education as objectives in the development of the nature park. There is thus a clear discrepancy between the regional perspective (from the

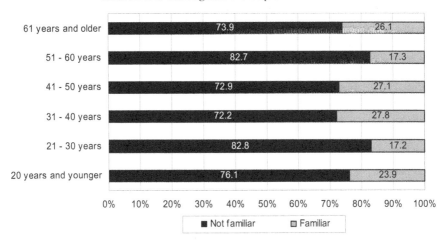

**Figure 5.9    People's familiarity with the nature park symbol (all municipalities, differentiated by age groups)**

*Source: Drafted by authors*

local population) and the outside perspective (that is, nature park concepts as defined by the VNÖ). Empirical studies, such as this one, contribute to the communication about this topic within the population. For example, in the municipality Puchenstuben alone, approximately one-half of the population was interviewed.

*Public Relations about the Nature Park*

Only a small percentage of the local population is familiar with the information material about the nature park. Such materials and media outlets include: brochures (37 per cent), information boards (15 per cent), printed media, local television or radio (11 per cent). Approximately 15 per cent of the interviewees – this is the second highest percentage – state that they do not know any public relation campaigns about the nature park. In Mitterbach, 26.7 per cent of the interviewed population is not familiar with any public relation campaigns. This is the highest percentage among all municipalities.

An important task of the management responsible for the Nature Park Region Ötscher-Tormäuer is to optimize the information-flow about the nature park to the local population. This can be achieved through an efficient, public, and engaging management campaign within the region (for example, emphasizing future plans, developments and projects of the nature park) as well as through promotions and self-projections to outside the region. Such a management campaign within the region would increase the perception the local population has about the number of educational activities in the nature park. The local population needs to be more informed about park events and activities and should be invited to participate in these activities. This could stimulate communication and cooperation within and between the local population and the nature park management.

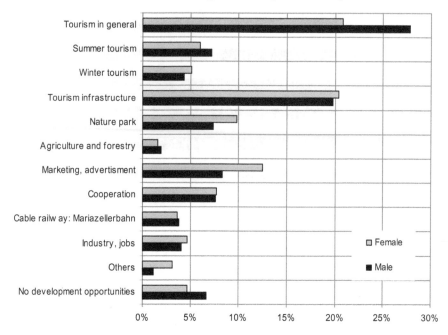

**Figure 5.10  Assessment of development opportunities (all municipalities, differentiated by gender)**

*Source: Drafted by authors*

As the results of the first study clearly documented, 76 per cent of the interviewees did not know the symbol of the nature park. These results did not change even when differentiated by age groups (see Figure 5.9). Interviewees seemed to dislike the symbol that showed two hikers and had been the same since the foundation of the park. Interviewees called the symbol 'old-fashioned', 'vague' and it reminded them of a popular TV character of a young girl, named 'Heidi', growing up in the Alps. For this reason, a competition for a new nature park symbol was conducted in 2003. The new symbol shows a group of brown bears, which inhabit the nature park. Although, brown bears are well known to the public, they are both sympathetic figures and a source of conflicts. Unfortunately, a survey about the familiarity and acceptance of this new symbol has yet to be conducted. It can be assumed that a strong integration of this symbol in an effective public management campaign would increase the identification with and the recognition of the nature park region among the local population.

*The Nature Park as a Potential for Growth?*

On average, 84 per cent of the interviewed population from the municipalities believes that the nature park can make a contribution to stimulating the economy. Almost 15 per cent of the interviewees think that the nature park will not improve the regional economy. While 88 per cent of Gaming states that the nature park will positively impact the regional economy, 'only' 78 per cent of Puchenstuben shares this opinion.

All interviewees explain this largely positive attitude towards the nature park with providing an impulse to the regional economy with increases in tourism (the nature park and what it has to offer increases the attractiveness and brings visitors) and associated with it a securing and financial support of restaurants and jobs.

The majority of the interviewees see the general future development possibilities of the Nature Park Region Ötscher-Tormäuer (see Figure 5.10) in the tourism industry (55 per cent overall). A comparison across different municipalities reveals that development potentials in the category 'tourism infrastructure' were more often mentioned in Puchenstuben, Mitterbach and Annaberg compared to the other two municipalities. The realization of deficits and the lack of infrastructure must be especially apparent in these two municipalities. Since nature conservation and the protection of the landscape are also objectives of the nature park, the planning of new infrastructure may possibly lead to conflicts of interest.

## Conclusions and Discussion: Strategies for an Integrative Management Exemplified by the Nature Park Region Ötscher-Tormäuer

The Nature Park Ötscher-Tormäuer is located in a structurally weak and underdeveloped region characterized by few new jobs, high out-migration, aging population, low employment, large number of commuters, high density of weekend homes, and so on. The local tourism marketing and regional managements (regional management 'Mostviertel' and LEADER+ Management 'Eisenwurzen') try to make the public aware of the name Ötscher-Tormäuer to be a mark of quality for a region. The cooperation between the tourism association and the LEADER+ management under the label 'Culture Park Eisenstrasse Ötscherland' is an indication of the local networking and marketing. In order to secure the future goal of developing an integrative management in the region, additional measures, experts believe, need to be introduced.

The following list prioritizes ten different issues that are believed to be relevant for a future regional management concept of the Nature Park Region Ötscher-Tormäuer:

1. *Easing the burden on the legal authority*: The legal authority of the Nature Park Ötscher-Tormäuer is, similar to most nature parks in Austria, an association. Its board includes the mayor and a second person from each municipality in the park. The multi-functional position of the mayor may prohibit him/her to contribute to the operational aspect of regional development (that is, nature park development), including effectiveness, project implementations and strategic development. The contribution of politicians should be rather seen as 'figure heads' in support of commonly (that is, across boundaries of municipalities) agreed upon development decisions. This requires the absolute necessity of an operational support structure, which takes responsibility for planning, development and project implementation. This leads to the next item:

2. *Management of the nature park*: The financial support over a three-year period (2003–2006) as part of an investment program for the Nature Park

Ötscher-Tormäuer has been an important basis for the development of the park. But this financial support ends at the beginning of 2006. The financial future of the park management is again uncertain and reminds of the time before the financial support started. Even the thirty hours per week position that was created during this time period is in jeopardy. If additional funding from different sources (for example, EU, federal, state, municipality) cannot be secured, then the successful and professional work of the park management cannot continue. Past experience has shown that marketing, the coordination of different interests and new ideas require professional and institutionalized supervision with adequate resources. This is the trend in professional regional development, which strongly emphasizes institutionalization and the moderation between interest groups. There are still many, so far unused opportunities that can be utilized in a coordinated fashion in order to secure the continuation of the park management with at least medium-term funding.

3. *Analog learning*: A stronger cooperation with other nature parks would be important, not only to strengthen the nature park idea, but also to stronger cooperation in areas of the nature park concept, including education and regional development that have been less developed until now (for example, specific educational activities for schools, field trips, and so on, as a coordinated offer from different nature parks with regional emphasis in Lower Austria). The first and most recent co-operations between the Nature Parks 'Buchenberg' and 'Lower Austrian Eisenwurzen' in 2005 give cause for optimism. In addition, a merger at least in some aspects of co-operation between all protected areas in the Eisenwurzen Region (including national parks), even across provincial boundaries would be extremely meaningful.

4. *Enlargement of the Nature Park Ötscher-Tormäuer*: The enlargement of the nature park in 2001 and 2002 can be categorized as a success of the nature park idea. However, it needs to be noted that mostly big landowners (for example Convent Lilienfeld) regarded this enlargement (of the size of the original enlargement plan, mostly in the municipality Annaberg) with fears and prejudices. Although the enlargement process of the nature park can be considered to have ended for now, it is important to keep up the dialogue and to try to harmonize the nature park with economic interests. This would be in accordance with nature park guidelines. In a region with a declining development at different levels, any useful option that could have a sustainable impact on development should be critically explored.

5. *Cooperation across municipality boundaries*: Coordination, communication, planning and securing of resources across administrative boundaries are still full of barriers. Regional protection could be a guide for a coordinated getting along with each other, even across provincial boundaries. Not only transnational programs of the European regional fund show the way to a coordinated getting along with each other. The protection of resources (for example, competition for the number of overnight stays) in a region that is characterized by many negative factors may not be a suitable survival strategy. Common responsibilities (for example, 'competence centers' in each participating municipality) must be developed, which are coordinated among

the municipalities and encourage tourists to mutual visits. The nature park management could take on a coordinating function.

6. *Coordinated development of tourism activities*: The Nature Park Region Ötscher-Tormäuer has many, partly segmented offers for tourism development. Coordinated marketing (that is, regional marketing) of individual activities that have until now been marketed by different municipalities, could be an important step towards a professional regional development. In the first place, this refers to tourism activities with respect to infrastructure and arrangements, but should also address marketing outside the region, especially in urban areas.

7. *Implementation of the four functions of a nature park*: As discussed above, primarily the functions of conservation, namely the protection of species, and recreation are present in the consciousness of the regional population. This is similar with experiences from other nature parks in Austria. The perceived implementation of the four objectives by the local population coincides very strongly with the actual measures and 'non-measures' in the region. Creating a togetherness of the four functions in nature parks on the same level seems to be primarily a challenge in terms of educating and raising of people's awareness. This is not only true for the affected population of nature park regions. The coordination and implementation of the objectives – protection, recreation, education and regional development – can be seen to be a very important goal for the Nature Park Ötscher-Tormäuer.

8. *Coordinating a set of measures by the EU*: As mentioned earlier, the larger study area (Eisenwurzen) is already the recipient of comprehensive regional financial support from the European Union (at least until the end of 2006). There is a need to establish measures of a regional management across state boundaries or to continue the coordination with different financial support after 2007. Although, the effects of state specific measures are encouraging (for example LEADER+ management in Ybbsitz/Lower Austria), the area of responsibility is also very locally defined. In general, there needs to be a stronger re-orientation from the side of the states. Even if it is important, in principal, to serve as the local spokesperson in his/her region, it would be more important to establish development measures together with other regions. A 'Europe of Regions' focuses on large regions and not on small regional development concepts. The region 'Eisenwurzen' offers numerous points of contacts for already existing projects that are partly stretching across state boundaries. The European Union also increasingly offers supporting measures, including NATURA 2000, and the Alpine Convention.

9. *Conceptual versus reactive*: Area and regional management in the region Ötscher-Tormäuer must be actively working together at a conceptual level. A nature park management has the task to influence the landscape in a forward-looking way. To proactively anticipate new trends in regional development instead of passively reacting to these trends must be seen as a responsibility of the management. Today, area management with respect to regional protection cannot be viewed any more as unimportant. Not only because of a declining employment in agriculture and forestry, but also because of a gradual decline in the size of a cultural landscape that has been used for centuries, is a change

of mind necessary. The discussion surrounds the preservation or the dissolving of a cultural landscape that has grown over centuries. Area management combined with measures of a regional management allows altering the use of newly available land for the purpose of integrative regional development. Area management with respect to regional protection is an alternative measure to the intensive use of large areas, such as golf and ski tourism.

10. *Ecological, social, and economical sustainability*: Without flogging to death the term 'sustainability' in this context, it needs to be pointed out that protected areas can be considered a classic example for a socially acceptable, resources securing development with economic profit for each region. This debate has been paid little attention to in the Ötscher-Tormäuer region. With a – similar to all regional development programs, including Lower Austria or the European Union – progressive decentralized concentration of funding measures for regional centers, the functionality of regions will become further important (see Amt der Niederösterreichischen Landesregierung 2004). There needs to be a push for an independent profile, balancing economy, sociology, and ecology under the guidelines of sustainable development. Nature parks, in general, and those close to urban agglomerations, in particular, can already now take over this role. Area management, in the context of regional protection, ultimately serves the function of a social, super-regional collective good with immediate benefits for the region in the sense of an active value added.

When looking into the future it needs to be noted that for nature parks in Austria – in order to overcome the ambivalence problem (conflict between environmental protection versus other uses) – attention must be paid to the goal of equal coexistence of protection, recreation, education, and regional development. Any claim for the exemplary nature of this model is lost if education and regional development are not successfully integrated into the concepts of nature parks and put into practice and if the participation from the regional population and activists is not secured and not presented to the 'outside'. Model and reality are still very much apart from each other.

The role the regional population and regional activists play with respect to their attitude towards the nature park has generally been unclear up to now. Not only contradictions (for example, economy versus ecology) can rarely be dissolved in participatory networks, but also especially business people (as shown in this chapter using the example of big landowners) express vehement opposition to extensive protected areas. Widespread reservations exist that communication will slow down or even prevent processes. On the other hand there is an involvement of activists that is not politically legitimized, embodied in a law, or possess the authority to act. The selective perception of regional activists thus blocks nature park regions frequently in their abilities to act. The contradiction of economic interests and goods (individual interests) opposite collective interests and goods (social interests) becomes clearly visible with nature parks. The lack of organizational structures in the form of nature park managements also makes the coordination of regional activists almost impossible.

The collaboration between the different levels of decision-making and instruments was not sufficiently discussed with this case study. European, national, regional, and municipal funding instruments and institutions do not always coordinate their activities. Parallel structures are therefore typical for the implementation of strategies for regional protection. The necessary competences for project implementation are mostly perceived differently. Down-up processes, that is, top down approaches that are further designed as bottom-up, which are also typical for strategies in protected areas, have been insufficiently dealt with at both theoretical and practical levels.

In addition, the question about collective implementation of resource management (for example, in regional protection) is still unclear in many areas. In contrast to the regional-economical development, where already a more comprehensive discussion about forms of 'home rule' and approaches to regional governance has been conducted, public goods (for example, national parks) have until now for the most part been ignored from this discourse. There exists a need for an additional motivation to commit to collective goods beyond a cost-benefit calculation (see Fürst et al. 2005). Regional planning approaches, which focus on image, regional identity and quality of life, will therefore be more significant in the future.

In conclusion, it should be noted that this case study not only demonstrates the possibilities that are associated with the concepts of nature parks in Austria. But it also primarily outlines the future room for maneuvers and necessities, which for the implementation of strategies of Austrian nature parks are still not clear at this moment.

## References

Amt der Niederösterreichischen Landesregierung (ed.) (2004), *Landesentwicklungs konzept* (Amt der Niederösterreichischen Landesregierung, St. Pölten).

Chape S., Blythe, S., Fish, L., Fox, P. and Spalding, M. (2003), *2003 United Nations List of Protected Areas, IUCN Gland, Switzerland and Cambridge, UK and UNEP-WCMC* (Cambridge, UK).

CIPRA-Österreich (ed.) (2002), *Wer hat Angst vor Schutzgebieten? Schutzgebiete als Chance für die Region* (Wien).

ETB – Edinger Tourismusberatung Gmbh (2002), *Freizeittouristisches Naturpark-Entwicklungskonzept, Endbericht* (Wien).

EUROPARC and IUCN (1999):, *Richtlinien für Management-Kategorien von Schutzgebieten. Interpretation und Anwendung der Management-Kategorien für Schutzgebiete in Europa* (Grafenau).

FNNPE – Föderation der Natur- und Nationalparke Europas (ed.) (1994), *Europarc. Proceedings of the 1993 FNNPE general assembly and symposium on protected area systems – the European experience* (Norfolk).

Fürst, D., Lahner, M. and Pollermann, K. (2005), 'Regional Governance bei Gemeinschaftsgütern des Ressourcenschutzes: das Beispiel Biosphärenreservate', in *Raumforschung und Raumordnung* 5, 330–9.

Hammer, T. (Hrsg.) (2003), *Großschutzgebiete – Instrumente nachhaltiger Entwicklung* (München).

Handler, F. (2000), 'Naturparke: Landschaft erleben – Natur begreifen', in OÖ. Akademie für Umwelt und Natur (ed.): *Naturspektakel oder Sanfter Tourismus* 43–8, (Linz).

Heintel, M. (2005), *Regionalmanagement in Österreich – Professionalisierung und Lernorientierung*, Institut für Geographie und Regionalforschung (Wien).

Heintel, M. and Weixlbaumer, N. (1998), 'Entwicklungsregion OÖ Eisenwurzen – Hintergründe und Ergebnisse einer sozialgeographischen Langzeitstudie zur OÖ Eisenstraße und Landesausstellung 1998 "Land der Hämmer"', in: Wohlschlägl, H. (ed.): *Forschungsberichte – Geographie Wien*, Institut für Geographie und Regionalforschung, 37–55 (Wien).

Mose, I. and Weixlbaumer, N. (eds) (2002), *Naturschutz: Großschutzgebiete und Regionalentwicklung* (Academia, Sankt Augustin).

Mose, I. and Weixlbaumer, N. (2003), 'Grossschutzgebiete als Motoren einer nachhaltigen Regionalentwicklung? Erfahrungen mit ausgewählten Schutzgebieten in Europa', in Hammer T. (ed.) *Großschutzgebiete – Instrumente nachhaltiger Entwicklung* (München), 35–95.

ÖGNU and Wolkinger, F. (eds) (1996), *Natur- und Nationalparks in Österreich* (Umweltdachverband ÖGNU, Wien).

Reusswig, F. (2004), 'Naturschutz und Naturbilder in verschiedenen Lebensstilgruppen', in Serbser, W., Inhetveen, H. and Reusswig, F. (eds): *Land – Natur – Konsum. Bilder und Konzeptionen im humanökologischen Diskurs*, München, 143–76.

Scherl, F., Tome, A., Broili, L., Perco, F. and Perco, F. (1989), *Parco Naturale delle Prealpi Carniche. Piano di conservazione e sviluppo. D.6: Norme per l'esecuzione del piano* (Trieste).

VNÖ (ed.) 2005, *Österreichische Naturparke. „Natur erleben – Natur begreifen", 10 Jahre Verband der Naturparke Österreichs*, Broschüre (Graz).

Weixlbaumer, N. (2005a), '„Naturparke" – Sensible Instrumente nachhaltiger Landschaftsentwicklung. Eine Gegenüberstellung der Gebietsschutzpolitik Österreichs und Kanadas', in *Mitteilungen der Österreichischen Geographischen Gesellschaft* 147: 34, 67–100.

Weixlbaumer, N. (2005b), 'Zum Mensch-Natur-Verhältnis – Naturparke als Innovationsfaktoren für Ländliche Räume', in *Alpine Raumordnung* 26, 7–18.

Zollner, D. and Jungmeier, M. (2003), *Kulturlandschaften österreichischer Naturparke*, Studie im Auftrag des Verbandes der Naturparke Österreichs, Bearbeitung: E.C.O. – Institut für Ökologie GmbH, Klagenfurt, 35.

# Chapter 6

# Hohe Tauern National Park: A Model for Protected Areas in the Alps?

Ingo Mose

## Introduction

The Hohe Tauern National Park holds a special position among the large protected areas in the Austrian Alpine region and of the Eastern Alps as a whole. It is the first one to include areas of three Austrian *Länder* (states), namely Carinthia, Salzburg and Tyrol. Among the largest projects in the history of spatial planning in Austria, its realization was for many decades marked by complex and long-lasting conflicts of interest. In the early 1990s, however, the nature reserve project was officially completed, and today it is the largest national park in the Alps and Central Europe. Therefore, its successful installation is undisputedly considered an enormous achievement of nature conservation (see Hasslacher 1999, 318). In addition to the mentioned attributes, the Hohe Tauern National Park is quoted as a model and even an 'exemplary piece' of Alpine area protection policy (see Hasslacher 1998, Hasslacher 1999). Thus, the nature reserve is put into the context of an important debate on principles, touching on both the fields of regional planning and of regional development policy, which in Austria are closely connected (see Schindegger 1999, 33). The central question is how to combine the protection of natural and cultural landscapes of ecological, cultural or esthetical value with the economic and socio-cultural development of mostly peripheral rural areas.

This question remains a constant and great challenge considering the ongoing regional structural change, which in the mountainous region of the Alpine area is characterized by far-reaching economic, socio-cultural and ecological problems (see Bätzing 1999). The example of the Hohe Tauern National Park demonstrates how the active involvement of private and civil society actors in decision-making processes, the consistent use of different funding programmes, the implementation of new regional institutions and the increasing co-operation of the local authorities involved successively contribute to elaborating action perspectives designed to combine area protection with regional development. At the same time, however, the example illustrates the deficits and problems deriving from such approaches.

## The National Park as a 'Lucky Chance' for Alpine Area Protection

Established in the years 1981, 1984 and 1991, the Hohe Tauern National Park covers large areas of the mountain range of the same name in the Austrian Central Alps. Its

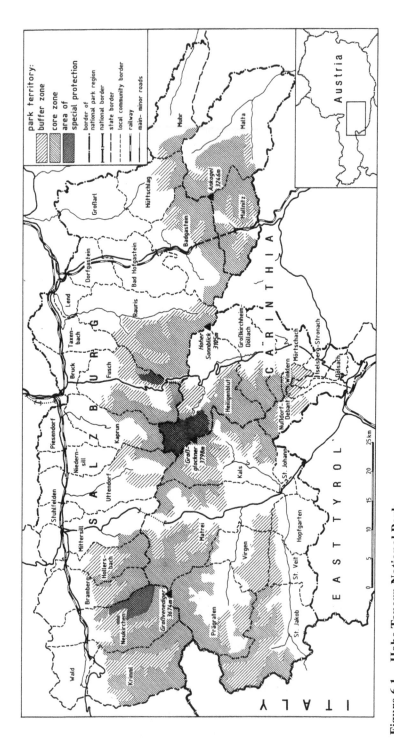

**Figure 6.1  Hohe Tauern National Park**
*Source: Drafted by author after Hohe Tauern National Park Authority*

most important parts are the Venediger Group, the Glockner Group and the Ankogel Group. The Hohe Tauern National Park extends over about 100 km from the Gerlos Pass in the west to the Mur valley in the east and covers 1,786 sq. km (see Figure 6.1). The park area is characterized by a cultural landscape which has developed over millennia and is traditionally used for mountain farming. In addition, the park includes extensive mountain forests, numerous unspoilt streams and large areas of Alpine wilderness characterized by permanent glacial ice cover and some of the highest summits of the Eastern Alps, for example the Grossglockner (3,876 m) and the Grossvenediger (3,674m).

Except for a few cases of seasonal settlement, for example on Alpine meadows and huts, the area of the national park is uninhabited. Approximately 58,000 people live in the area surrounding the national park, including the adjacent local communities which contribute varying sizes of territory to the park.

Efforts to establish a national park in the Hohe Tauern area date back to the early twentieth century. However, it was not until 1971 that the federal heads of the provincial governments of Carinthia, Salzburg and Tyrol undersigned the *Heiligenblut Agreement* to establish a multi-province national park. The objective of this agreement was to place this intact natural landscape under permanent protection.

After lengthy negotiations, arrangements regarding the park concept were made with about 1,100 landowners in the national park. Besides, there were several cases where the establishment of a three-province national park stood opposed to large-scale infrastructure development projects (the erection of hydro-electric power stations, projects for opening up the glaciers for skiing). Finally, the national park was officially founded in the Province of Carinthia in 1981, in Salzburg in 1984 and – considerably later – in Tyrol in 1991. Although the protected area is 30 per cent smaller than originally planned, with an area of 1,786 sq. km the Hohe Tauern National Park is the largest national park in Central Europe (see Figure 6.1).

The national park includes areas of the three provinces of Carinthia, Salzburg and Tyrol. Regarding protected areas it needs to be taken into account that in Austria nature protection policy primarily lies with the legal competence of the federal provinces of Austria. Three state National Park Laws (including the ordinances issued thereunder) therefore govern the establishment and management of the Hohe Tauern National Park. As a consequence, the park features three different administration units, including own personnel and budget, one based in each province.

The so-called National Park Council is the co-ordinating body of the three separate national park administrations. In 1997, the Hohe Tauern National Park Academy was established as a joint educational institution. Basic planning instrument for the state national park administrations is the respective national park plan which outlines the current status, overall objectives and specific actions for the future planning and development. The main tasks faced by the national park administrations include management of the natural environment, tourism and recreation, science and research, environmental education and public relations.

In an attempt to develop spatial planning concepts for the national park, a zoning concept has been introduced (see Figure 6.1). For achieving both nature protection and economic development (agriculture, tourism, energy industry and so on), the national park area was subdivided into zones which include landscapes modified by human impact and zones of largely unspoiled nature. This is reflected by the division of the national park into a buffer zone, a core zone and special protected areas. In general, the national park achievements of nature conservation are regarded as successful with respect to both the size of the protected area and the quality of the protected landscape.

A high level of protection is provided for the core zone and the special protected areas the status of which is equivalent to that of nature reserves. Thus several projects involving the use of countless untouched glacial stream ecosystems for producing energy were rejected. Impressive results of biotop protection and wildlife management are likewise the conservation (ibex and others) and the reintroduction (bearded vulture) of endangered species in the protected area (see Hasslacher 1999, 318).

The development of the protected area is, however, not finished yet. Ten years after its official founding, the park does not have full international recognition as a national park in accordance with category II of the International Union for Conservation of Nature and Natural Resources (IUCN) which all three national park administrations are determined to attain.

In July 2001, the Carinthian part of the Hohe Tauern National Park was granted international status as a national park under IUCN category II. To attain category II status, it was necessary to establish nature zones designated to free and natural development – that is not affected by human impact – of at least 75 per cent of the core zone and special protected areas and the preservation of the traditional cultural landscapes of at the most 25 per cent of the area. The same applies to the buffer zone with IUCN category V status (protected landscape). An important key to attain international status as national park was the restriction of hunting in the protected area which was necessary to meet the IUCN criteria.

The solution found is based on hunting leases by the national park administration and agreements with huntsmen on wildlife management according to the requirements of IUCN category II (see Kärntner Nationalparkfonds 2001). Against the background of the development in Carinthia, it is expected that the other two provinces will make progress in attaining national park status in accordance with the IUCN national park criteria.

## On the Road to Becoming a National Park Region

The finally successful implementation of spatial planning actions in the national park is based on a concept which was not yet developed when the national park planning began. The guiding principle of this concept is to protect and promote the protected area not only by developing a spatial zoning which keeps nature conservation, agriculture, tourism, etc. into account but also by implementing a regional development policy which aims to promote the surrounding area of the national park.

This includes compensation payments for (possible) income losses due to the fact that several large-scale energy and tourism projects were not realized due to the political decision to establish the national park. The necessary funds are provided in accordance with the agreement between the national government and the *Länder* governments on the co-operation in the field of protection and promotion of the national park which since 1982 provides a basis for allocating national park funds to the communities in the national park (see Hasslacher 1998, 11–12). In 1990 and again in 1994 – after the national park had been created in Tyrol – the co-operation between the federal government and the provinces was determined by state treaty. The main emphasis is on measures designed to preserve and promote Alpine agriculture as well as to develop sustainable tourism. The federal and provincial funds allocated for project-related measures are considerable and from 1982 until today an amount of approximately 50 million Euro has been provided.

The intended integration of area protection and regional development is based on the idea to make the development of the protected area compatible with the overall regional development planning beyond the actual protected area. To achieve this objective the adjacent area of the Hohe Tauern National Park has been defined as 'National Park Region' (see Fally 1994) (see Figure 6.1). It comprises 29 national park communities with land in the nature reserve as well as nine nearby or interspersed communities which do not contribute territory to the national park but are closely linked to the 'real' national park communities and, therefore, can be regarded as a common functional area (approx. 87,000 inhabitants).

Today, several concepts, development programmes and projects explicitly focus on the national park region (or parts of it) and are designed to strengthen the standard of living and economic foundation of the local population (see among others, Salzburger Institut für Raumordnung und Wohnen 1993a).

Particularly with regard to the economic development of the region, the role of the national park should not be underestimated. Outside agriculture, which is confronted with problems associated with mountain regions and today is increasingly dominated by part-time farming, only few industrial jobs could be created (namely contracting, wood processing, ski and snowboard production, etc.). Therefore, tourism and its related services soon became the most important economic sector. Due to the continuous development of winter sports, two-season tourism has increased in many communities. The number of overnight stays during the winter season considerable exceeds that of the summer season. The establishment of the national park has contributed to an increased quality of the tourism profile and potential of the involved communities in the sense of 'soft tourism'. Concrete measures include restructuring and new development of tourism in the national park area (see Mose 1988, Mose 2002). The example of the Glockner area illustrates the range of related infrastructure measures (see Figure 6.2).

These measures generate additional income and jobs in the region. A study assigned by the Salzburg National Park Fund shows that the tourism effects of the protected area are quite considerable in the accommodation and catering sector. Out of the total two million overnight stays in the vicinity of the protected area, 108,000 are directly induced by the national park and are also relevant for other sectors of the regional economy (retail trade, printing industry, credit institutes and so on) (see Österreichisches Institut für Wirtschaftsforschung 2000). The role of the national park administrations and services as employer is often ignored, even though they provide 71 jobs.

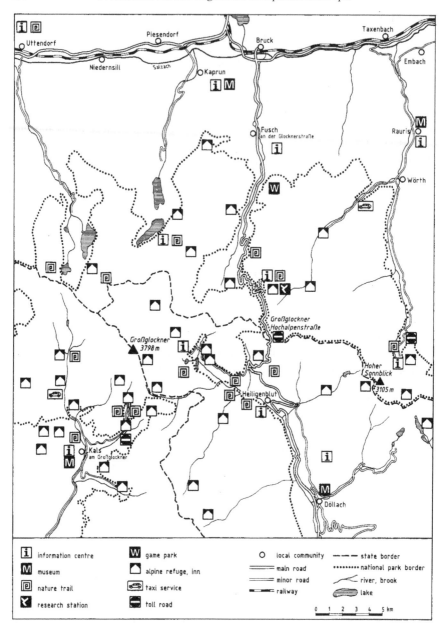

**Figure 6.2    Tourism infrastructure in the Großglockner area**
*Source: Drafted by author*

In the last decade, European regional development programmes, from which the region benefits since Austria joined the EU in 1995, were of particular importance for the initiation of regional development processes in the vicinity of the national park. From 1995 to 1999, nearly all national park communities were classified as objective 5b areas. After the reform of the structural funds these areas either take advantage of funding offered through the new objective 2 programme or have been classified as phasing-out areas (until end of 2006). Several regional management organizations provide services to optimize regional economic development and co-ordinate structural fund programmes in close contact with the regional economy. These organizations, however, operate on a larger scale and, as a result, are only partly related to the national park (see Mose 2003, 50pp.). From the actor perspective, the establishment of four Local Action Groups (LAGs) under the EU LEADER initiative in and around the national park, including the ARGE Hohe Tauern National Park in the Salzburg part of the protected area, is of considerable importance.

Also during the current LEADER+ funding period from 2000 to 2006, the LAGs have elaborated various projects in the areas of tourism, agriculture, trade, energy supply and culture. They are supported by LEADER managers. According to Fidlschuster (1998), the LEADER regions in the Hohe Tauern national park area rank among the most innovative ones in Austria. These activities illustrate the considerable activation of both the different regional actors and the regional population which correlates with a high acceptance of the protection area today. Especially in the initial phase, forms of informal active involvement of the local population, for example the establishment of the Future College Hohe Tauern National Park in the Salzburg national park communities, could be successfully used for the development of the national park region (for example preservation of historic buildings, cultural and educational activities). In addition, the TAURISCA association was set up in the province of Salzburg to promote various independent cultural initiatives in the national park region.

The influence of several NGOs such as the Austrian Alpine Club (OeAV) should also be mentioned. 21 per cent of the national park area is owned by the Alpine Club which has from the very beginning played a major role in the creation of the national park and the subsequent development into a national park region. This is also expressed by the initiation of different projects for which the Austrian Alpine Club is still responsible.

The outlined development approaches reflect the idea to place the development of the protected area in a comprehensive context, namely that of the National Park Region (see above). This idea has been adapted to several projects, including, among others, the regional brand 'Hohe Tauern National Park' which was developed in the framework of LEADER and is used as a common quality label by several agricultural, handicraft and gastronomic partner enterprises in the Salzburg part of the protected area. It was not until 2001 that the 'Holiday Region Hohe Tauern National Park ltd.' was founded by the Salzburg national park communities with the intention to generate greater publicity for the national park region as an international tourist destination.

## Prospects and Problems of the Previous Development

The initiated development process in the Hohe Tauern National Park area and its vicinity is in sharp contrast to previous action approaches in spatial planning and regional policy in the 1960s and 1970s which were primarily based on an interventionist top-down development approach, very often neglecting the endogenous potentials as well as the specific needs of the regions. In contrast, today's action approaches are characterized by a more de-centralized, regionally adapted development policy. This change is certainly an improvement because now specific regional interests can be articulated more strongly than before. The national park is an unusual experimental field for the implementation of these interests. Its function is not restricted to that of a protected area in a narrower sense as it increasingly plays the role of a regional development instrument. The realization of numerous innovative projects impressively proves the dynamic of the initiated development processes.

However, deficits and problems are apparent which diminish the supposed model character of the Hohe Tauern National Park. So far, at best selective scientific research has been done and further deepening and discussion is needed. Therefore, in the following only a few 'key questions' are presented.

An essential prerequisite for the successful implementation of the new development strategies is to ensure a broad acceptance by the local population. This is not unproblematic. It has to be noted that the active involvement of the population in local and regional development processes varies considerably from community to community as well as over time. While in the creation phase of the national park forms of active involvement of the population were achieved in some communities, the mobilization of such 'bottom-up'-initiatives was less successful in other communities. Today, the willingness to participate in concrete projects is rather unstable. Dependence on individual persons, the burden related to voluntary work, deficits regarding the transparency of the ongoing processes and the limited influence at local level might explain this tendency. This situation reveals the problems connected with the long-term embodiment of participative forms of planning. It is not yet foreseeable to what extent it might be possible to foster local participation processes in the framework of other initiatives, e.g. Agenda 21 which has been used as an succesful means of mobilization elsewhere (see Vilsmaier and Mose 2005). More recently in 6 communities of the Upper Pinzgau area the transdiciplinary research project 'LEBEN 2014' (2002–2005), conducted by the University of Natural Resources and Life Sciences Vienna and the University of Salzburg, has provided an interesting platform for local involvement. Aiming at a conceptual sketch of future development in the National Park Region, members of the local population and representatives of relevant institutions took part in an intensive discussion process regarding future scenarios for the region. As a result, a number of project ideas have been defined which ought to be implemented in the near future (see Freyer and Muhar 2006).

A further problem is the fact that it is difficult to ascertain which regional actors could function as 'regional institutions' with an explicit regional action perspective. A group of possible actors that could fulfil this requirement are the LAGs which were created in the framework of LEADER. Thus, it was the declared aim of the Community Initiative LEADER 'to support ... local action groups ... jointly devising a strategy and innovative measures for the development of a rural area on the scale of a local community' (see Europäische Kommission 1994, 20). This is an implicit reference to the action dimension of the 'regional level'. In fact, this dimension is

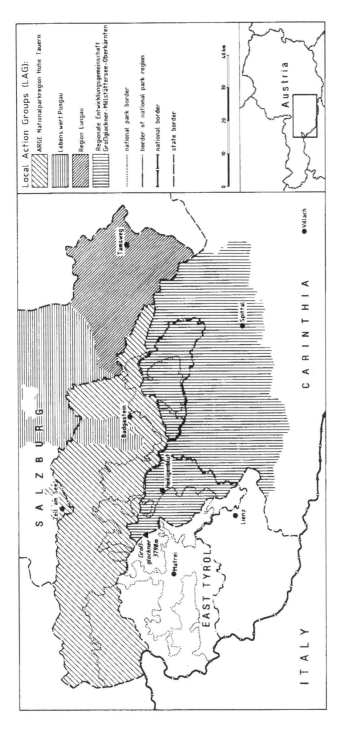

**Figure 6.3   Local Action Groups under the European Common Initiative LEADER+**

*Source: Drafted by author after www.leaderplus.at*

also reflected in the practical work of the LAGs. The regional brand 'Hohe Tauern National Park' in Salzburg is an example of this.

Even though efforts were and are being made to develop interactions between the individual LAGs, these are nevertheless restricted to their particular fields of activity. Besides, it has to be noted that the activities of the LAGs are limited to selected fields of action that are not free of sectoral particularist interests (agriculture, tourism and so on). Therefore, LEADER+ focuses on integrated and sustainable development concepts at the regional level (see Amt der Kärntner Landesregierung 2001). With the assistance of LEADER, it was obviously possible to develop successful action approaches for rural areas adjacent to the Hohe Tauern National Park which, however, are not yet a concept for the development of the entire national park region as defined by Fally (1994).

The regional management initiatives which were created after joining the European Union are even less appropriate than the LAGs to serve as an institutional link for the national park regions. The mentioned institutions have in common that their field of activity in the national park region is very limited. They are related to other (administrative) regions, and their tasks consist, above all, in promoting regional economic development as a whole. Nevertheless, the regional management initiatives are important actors in the national park region.

The national park administrations were also not able to compensate the lack of a 'regional institution'. By definition, these are institutions at provincial level, which clearly indicates their sphere of activity. Their responsibility is the administration and management of the clearly defined territories of the respective protected areas. However, numerous activities of the administrations directly or indirectly influence the development in the national park region, especially in the field of tourism. A close co-ordination of measures in both the protected area and the surrounding area is important (for example development for traffic infrastructure, information facilities). This, however, does not mean that the national park administrations have a specific mandate to engage in regional policy. In addition, the national park administrations cannot and will not speak in 'one voice' for the national park or the national park region. It is not yet known if it will be possible to set up a joint administration for the cross-provincial protected area, even though a common administrational structure would provide numerous advantages, particularly as the national park is designed to be perceived as a coherent protected area both by the local population and visitors from outside. Respective considerations are, however, in conflict with the well-established autonomy of the involved provinces.

Finally, it is not certain if it is possible to combine all relevant initiatives in a common model or common regional development concept. In fact, a national park model or development concept for the national park region that could serve as a common guideline for regional development is not yet available. In 1995, the National Park Council decided to adopt common 'guidelines' which, however, only indicate objectives for the management of the protected area in the narrower sense and do not include statements regarding the development of the larger national park region (see Nationalparkrat 1995). The 'Development and promotion concept for the National Park Hohe Tauern in Salzburg', established in 1993 for the national park administration (see Salzburger Institut für Raumordnung und Wohnen 1993a,

Salzburger Institut für Raumordnung und Wohnen 1993b) meets this demand only to a certain extent since it merely consists of a catalogue of potential – very general – objectives and measures. After more than ten years it is still not clear how the relevant institutions – especially those engaged in regional planning and regional economic development – could implement the concept as a guideline for an 'overall concept' for the (Salzburg) national park region.

This deficit in spatial planning and regional policy for the entire national park region corresponds to the deplorable state of affairs at the level of partial areas. In almost all national park areas of Carinthia, Salzburg and Tyrol regional planning has not yet managed to update or develop the legally required development programmes for the respective partial areas (see Hasslacher 1999, 321–22).

The example of the Salzburg districts that contribute territory to the national park shows how outdated the planning data are. The relevant development programmes for supra-local spatial planning date from the 1970s and 1980s (see Salzburger Institut für Raumordnung und Wohnen 1993a, 196pp.); the 'Regional Programme' for the Lungau district, passed in 2000, is an exception (see Kals and Schönegger 2000). This lack of clear regional planning objectives for the future development in the area around the national park is all the more disappointing given that the indicated regional management institutions have been established since the mid-1990s and it would be useful to co-ordinate their work with the objectives of regional planning.

Both the districts around the national park and the planning area of the entire national park region still lack a regional development concept designed to provide an orientation for the future development of this area. There is no legal basis for regional planning in the provinces involved that provides the possibility to implement such an instrument for large-scale planning or development beyond the boundaries of the federal provinces. However, such an instrument would be useful because the previous planning and development concepts at partial area or sectoral level do not explicitly or only partially address a 'regional perspective'.

## Summary and Outlook

In several respects the Hohe Tauern National Park can be regarded as an interesting 'test case' for the development of integrated forms of area conservation policy in the Alpine area. This applies especially to the implementation of regionally adapted strategies and instruments of regional development which according to Bätzing (1999) are an indispensable prerequisite for finding adequate solutions to the increasingly specific development problems in the Alpes at regional level. Experiences with the development in the Hohe Tauern national park region show a heterogeneous and conflicting picture of the implementation of respective action strategies. Some important results of the above analysis can be summarized as follows.

The spatial planning quality of the implementation of the national park in all three provinces is largely undisputed. On the basis of a spatially differentiated zoning concept, it was possible to achieve a sustainable balance between the interests of nature conservation on the one hand and economic interests on the other hand. Even though the international recognition of the entire national park is still lacking, there is no doubt that its implementation is an enormous improvement for Alpine area conservation. Especially the fact that vast areas of the Hohe Tauern could be protected from large-scale infrastructure projects cannot be valued highly enough.

As a result of the establishment of the national park, numerous regional projects and initiatives were initiated which, in sum, had a very positive effect on the development of the national park region. Various measures in the fields of agriculture, tourism and handicraft, but also in the educational and cultural sector can be mentioned as examples. There is evidence that economic impulses in the form of income and jobs result directly or indirectly from the creation of the national park although these have not been measured fully yet.

Less clear, however, is to what extent the initiated activation can be regarded as an expression of regional action as intended with the construction of the national park region. So far, the development is mainly characterized by the fact that different approaches for more regionally oriented action in spatial planning and regional development policy (regional management, LAGs, tourism marketing and so on) (still) are only additively related and their activities have not yet been sufficiently integrated. This applies to the co-ordination within the provinces involved and beyond the provincial boundaries.

These findings confirm the observation that the competition between different actors – especially of the communities but also other institutions – is still a strong, though not the only, motor for regional development in most partial areas of the national park region. Examples for this are the controversial attitudes towards additional tourism infrastructure planning. The conflict between the further development of a national park-oriented tourism on the one hand, and the attempts to realize large-scale tourism projects that have no connection to the protected area on the other hand, may serve as an example (see Hasslacher 1999; Mose 2002). This is inconsistent with the declared aim of regionally co-ordinated action which requires a stronger and long-term co-ordination and co-operation. The formulation of a regional overall concept or regional development concept for the total area of the national park region might be a suitable basis for achieving this objective. Although various actors postulate the necessity to promote an economically acceptable, environmentally friendly and sustainable regional development (see among others, Salzburger Institut für Raumordnung und Wohnen 1993a, 2), a framework for the national park region in which these objectives are embedded is still lacking.

These findings need especially to be considered under the aspect that the establishment of a high-ranking large-scale protected area such as the national park offers enormous prospects. Kals (1997, 81) is right in stating:

> Protected areas are too valuable to serve as a tranquillizer for the environmental conscience .... Therefore, the aim must be a real integration of the protected areas into the overall development of a region. Protected areas can therefore function as centres to create an identity for model regions, including the subsequent consistent orientation of the total area towards the 'regional idea' [emphasis in the original].

This picture corresponds to further ideas to consider protected areas as 'recipes' or even as 'model landscapes' for the development of rural areas in general (see

Weixlbaumer 1998b). The ongoing discussion in regional studies, however, shows that 'new' types of protected areas are considered to be more appropriate which are explicitly directed to integrate conservation and development of a region in the sense of a dynamic-innovative approach. At present, this characteristic is mainly attributed to biosphere reserves (or biosphere parks); in some studies, the possible role of regional parks and nature parks is assessed similarly (see chapter 1).

In this context, the Hohe Tauern National Park obviously represents an 'interim model', consisting of static as well as dynamic area protection. Thus, the national park explicitly functions as an instrument of both area protection and regional development. An evaluation of the previous development of the Hohe Tauern National Park should be based on a careful consideration of the undoubtedly achieved great successes and the apparent deficits of the course adopted. Hasslacher (1998), for example, repeatedly called for a careful assessment. According to him, 'a successful development characterized by a European-wide "aha" association and an unmistakable model character ... has not yet been achieved' (Hasslacher 1998, 12).

In fact, several expectations placed in the national park and the national park region have not been fulfilled, especially with regard to the integration of the protected area in important international political action approaches. Until today, for example, none of the national park communities has become a member in the community network 'Alliance in the Alps' which aims to realize the goals of the Alpine Convention and its protocols in the Alpine area at community level.

In addition, it is still unclear how the Hohe Tauern National Park could function as a 'catalyst' for a large-scale interregional area protection network in the main crest of the Alps whose potential importance for the Alpine area protection cannot be emphasized strongly enough. However, the involvement in internationally binding co-operations does not make sense as long as the deficits with respect to a sustainable conceptual co-ordination among all relevant actors in the national park region persist.

In the future more than today, the actors involved will have to face the challenges of protected area development and to see the national park as what it already is: not (yet) a model case, but undoubtedly a great hope for Alpine area protection policy!

## References

Amt der Kärntner Landesregierung (2001), *Die neuen EU-Förderprogramme der Orts- und Regionalentwicklung bis 2006* (Klagenfurt).

Bätzing, W. (1999), 'Die Alpen im Spannungsfeld der europäischen Raumordnungspolitik', *Raumforschung und Raumordnung* 1, 3–13.

Blotevogel, H.H. (1996), 'Auf dem Wege zu einer "Theorie der Regionalität": Die Region als Forschungsobjekt der Geographie', in G. Brunn (ed.): *Region und Regionsbildung in Europa* (Baden-Baden), 44–68.

Danielzyk, R. (1998), *Zur Neuorientierung der Regionalforschung* (Oldenburg).

Eder, U. (1998), *Naturschutz grenzenlos – oberstes Gebot der Stunde oder Utopie? Ein humangeographischer Beitrag zur Diskussion der Vernetzung alpiner Schutzgebiete am Beispiel des geplanten INTERREG II-Projektes 'Alpensteig' im Naturschutzverbund Nationalpark Hohe Tauern, Naturpark Rieserferner-Ahrn und Ruhegebiet Zillertaler Hauptkamm.* Unveröffentlichte Diplomarbeit Universität Wien (Wien).

Europäische Kommission (1994), *Leitfaden der Gemeinschaftsinitiativen 1994–1999.* 1st edition (Luxemburg).

Fally, W. (1994), 'Die Region "Nationalpark und Vorfeld" als Funktionsraum', in Roland Floirmaier (ed.), *Umdenken. Zehn Jahre Nationalpark Hohe Tauern in Salzburg – eine Bestandsaufnahme* (Salzburg), 107–20.

Fidlschuster, L. (1998), 'Nationalpark Hohe Tauern – ein Katalysator für innovative LEADER-Projekte', *Raum* 32, 13–14.

Freyer, B. and Muhar, A. (Hrsg.) (2006), *Transdisziplinäre Kooperation in der universitären Ausbildung. Die Fallstudie 'Leben 2014' in der Nationalparkregion Hohe Tauern/Oberpinzgau* (Wien).

Hasslacher, P. (1998), 'Nationalpark Hohe Tauern: Ein Lehrstück alpiner Raumordnung', *Raum* 32, 10–12.

Hasslacher, P. (1999), 'Die Entwicklung des Nationalparks Hohe Tauern – eine raumordnungspolitische Zwischenbilanz', in Gerhild Weber (ed.), *RaumPlanerStoff* (Wien), 317–26.

Hasslacher, P. (2000), 'Schutzgebietsverbund Alpenhauptkamm', in Salzburger Nationalparkfonds (ed.), *Das Krimmler Tauernhaus und seine Umgebung in Geschichte und Gegenwart.* (Neukirchen am Großvenediger), 87–94.

Kals, R. (1997), 'Schutzgebietsmanagement als integrierter Bestandteil der Regionalentwicklung', *Alpine Raumordnung* 14, 80–84.

Kals, R. and Schönegger, C. (2000), 'Neue Regionalplanung im Land Salzburg – Beispiel Lungau (Bezirk Tamsweg)', *Ländlicher Raum* 13:2, 7–10.

Kärntner Nationalparkfonds (ed.) (2001), *Nationalpark Hohe Tauern in Kärnten – Der Weg zur internationalen Anerkennung* (Großkirchheim).

Mose, I. (1988), *Sanfter Tourismus im Nationalpark Hohe Tauern* (Vechta).

Mose, I. (2002), 'Nationalpark Hohe Tauern – Tourismusentwicklung und Schutzgebietsplanung im Alpenraum', in J. Borghardt, L. Meltzer, S. Roeder, W. Scholz and Wüstenberg, A. (eds), *ReiseRäume* (Dortmund), 183–198.

Mose, I. and Weixlbaumer, N. (2003), 'Großschutzgebiete als Motoren einer nachhaltigen Regionalentwicklung? Erfahrungen mit ausgewählten Schutzgebieten in Europa', in T. Hammer (Hrsg.), *Großschutzgebiete – Instrumente nachhaltiger Entwicklung.* (München), 35–96.

Nationalparkrat Hohe Tauern (1995), 'Nationalpark Hohe Tauern – Leitbild', *Beschluß des Nationalparkrates vom 10.8.1995* (Salzburg).

Nationalparkrat Hohe Tauern (1998), *Nationalpark Hohe Tauern. Offizielle Information.* (Matrei in Osttirol).

Österreichisches Institut für Wirtschaftsforschung (2000), *Ökonomische Effekte des Nationalparks 'Hohe Tauern' auf die Wirtschaft des Bundeslandes Salzburg* (Wien).

Salzburger Institut für Raumordnung und Wohnen (1993a), *Entwicklungs- und Förderungskonzept für die Nationalpark Hohe Tauern Vorfeldregion im Bundesland Salzburg. Teil I: Bestandsaufnahme und Analyse* (Salzburg).

Salzburger Institut für Raumordnung und Wohnen (1993b), *Entwicklungs- und Förderungskonzept für die Nationalpark Hohe Tauern Vorfeldregion im Bundesland Salzburg. Teil II: Generelles entwicklungspolitisches Ziele-, Maßnahmen- und Förderungskonzept* (Salzburg).

Schindegger, F. (1999), *Raum. Planung. Politik. Ein Handbuch zur Raumplanung in Österreich* (Wien).

Vilsmaier, U. and I. Mose (2005), 'The Implementation of the National Park Idea in Society – The Role of Agenda 21 Processes'. In: Hohe Tauern National Park Council (ed.) (2005), *Conference Volume – 3$^{rd}$ Symposion of the Hohe Tauern National Park for Research in Protected Areas* (Matrei in Osttirol), 239–242.

Weixlbaumer, N. (1998a), *Gebietsschutz in Europa: Konzeption – Perzeption – Akzeptanz* (Wien).

Weixlbaumer, N. (1998b): 'Schutzgebiete. Ein "Rezept" für die aktive Sanierung ländlicher Räume?', *geographie heute* 19:164, 30–3.

Chapter 7

# Regional Development and the French National Parks: The Case of the Vanoise National Park

Isabelle Mauz

## Introduction

Many people think that the French national parks are exclusively dedicated to environmental protection. Their regulations, which forbid hunting for instance, are in fact fairly strict. But their reputation is exaggerated and it is not rare to hear that permissible activities, such as grazing, are forbidden. It is also commonly thought that the national parks have nothing to do with local development, contrary to the regional nature parks. It is the latter that are mentioned, in France, in discussions on coordinating protection and development and they are also thought to be driving forces in sustainable development. The discussion in this chapter will show that the situation is much more complex and that the attention paid to local development in the national parks has varied considerably from their inception to the present time.

The Vanoise National Park was selected as a case study for two reasons. First, it is the oldest of the French national parks and its history had major influence on the 1960 law that established the national parks as well as on the parks subsequently created.[1] Secondly, the history of the park has been the subject of a number of studies (see Mauz 2003; Laslaz 2004; Mauz 2005) and is therefore better known than that of the later parks.

We distinguish five periods after the initial time when the possibility of creating a national park in the high Savoy valleys started to be seriously discussed. The first was of course the project period. We will see that one of the projects explicitly targeted the development of the local economy and significantly contributed to widespread acceptance of the Vanoise park and of national parks in general. The second step consisted primarily of drafting the founding documents, which structured the park into the central and peripheral zones, determined that protection was the principal mission and relegated the inhabited zones and local development to a secondary role.

---

1  France has six other national parks, namely Port-Cros, created on 14 December 1963, the Pyrenees, created on 23 March 1967, the Cévennes, created on 2 September 1970, the Écrins, created on 27 March 1973, the Mercantour, created on 18 August 1979, and Guadeloupe, created on 20 February 1989 (see Figure 7.1).

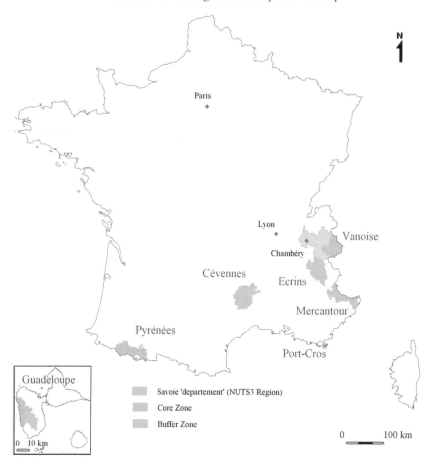

**Figure 7.1    The French national parks**
*Source: The Vanoise National Park Authority*

The third step started with the effective creation, on site, of a territory that until then had existed only on paper. The immediate result was a double retreat, geographically into the central zone of the park and functionally with a focus on the protection mission. It was only when the park was firmly established and its presence recognized that the park officials could open up to the outside world and establish relations with the actors involved in local development (step four). Finally, step five started with the report by MP Jean-Pierre Giran on the national parks, submitted in June 2003. Since then, a draft law has been prepared that is largely based on the conclusions of the report. One of the goals is to enhance the influence of local elected officials in the management of national parks and to provide the public structures running the parks with the means to better combine protection and development.

## A National Park Project Targeting Local Development

At the end of the 1950s, there were two major projects for the creation of a national park in Savoy. The primary goal of the first project to be presented was to help the ibex recolonize the high valleys, starting from the Gran Paradiso National Park where the animal had already been protected for many years. The proposed boundaries were drawn in view of establishing a zone next to the Italian park and including all the French areas where the ibex had been observed. The park was also sized to meet the needs of a large ibex population. The project planned to hire many employees in a strictly hierarchical structure, where the essential job of the employees would be to monitor the animals, keep a close eye on hunters and impose heavy penalties on poachers. The local population was not completely ignored. The project noted that hunting zones must not be too severely limited to avoid negative reactions and that the presence of the local population did have its advantages. In this way people with excellent knowledge of the area as well as local inhabitants could be rapidly hired and became excellent guards. The virtually finalized project was promoted by Marcel Couturier, a surgeon from Grenoble, who was also a well-known naturalist, excellent mountaineer and inveterate hunter.

The other project was very different. It was the product of a boundless admiration and an equally determined refusal. The admiration was for the natural beauty of the Alps, but also for the traditional communities that were viewed as the sole antidote to the large cities considered the cause of the decadence and despair of modern man. The project refused to see the communities slowly lose their inhabitants and the Alps be taken over by hydraulics engineers and the promoters of tourist resorts. A national park was needed to stop the emigration of the mountain population and muzzle the projects of outside promoters. Homes were to be renovated, the roads made safer, the trades and pastoral farming encouraged and developed. The limited installation of rich urban populations was to contribute to providing the local people with the resources required to stay. Finally, mountain schools were to be created, where certain young urban dwellers would be removed from the negative influence of the cities and exposed to "mountain life". Plants and animals, both domestic and wild, were also taken into account. All of mountain life had to be preserved, but above all that of the inhabitants who were considered the basis of all possible development. This second project was the brainchild of a young man from the Vosges, Gilbert André, fascinated by the Alps since his childhood and who, after having walked many a trail, stopped one day in June 1956 in Bonneval-sur-Arc, the most remote village in the Haute-Maurienne valley, where he devoted himself to the realization of his idea of a national park.

Though virtually everything differed in their projects, Marcel Couturier and Gilbert André both proclaimed, at the same place and time, that it was absolutely necessary to create a national park. But what they meant by the term "national park" was of course very different. Marcel Couturier designed a park tailored to the ibex and entirely devoted to its protection, whereas Gilbert André imagined a much vaster park, including the valley bottoms because the goal was to stimulate the villages.

The naturalist and humanist projects did not curry favour among the same groups. In Savoy, the idea of protecting the ibex appealed to mountaineers and big-game hunters from the surrounding cities, Chambéry, Grenoble and Lyons. The project to support and modernize crop and animal farming, develop the trades, improve roads and make them safer, in short, enable the mountain populations to "stay put» was warmly welcomed by local elected officials confronted with a declining population in the high valleys. The Haute-Maurienne valley in particular continued to lose inhabitants and the future of a village such as Bonneval-sur-Arc, which suffered catastrophic flooding in June 1957, appeared bleak. The national park as presented by Gilbert André was viewed as an original development technique. Finally a solution had been found to halt the slow decline and the despair of the population. It happened that at the end of the 1950s, Savoy could boast a number of influential politicians on the national scale. Pierre Dumas (1924–2004) was elected MP in 1958 and served as a minister from 1962 to 1969. Joseph Fontanet (1921–1980), was elected MP in 1956, vice-president then president of the departmental council, and was also called to Paris a number of times as a minister. At the other end of the political spectrum, Pierre Cot (1895–1977) was a departmental councillor, former MP and minister under Léon Blum. The three men did agree on one point: a national park represented a chance to save the mountain regions from their dilemma. Between the two projects for the park, they did not hesitate and clearly preferred the idea of the park serving local development. It was in development terms that the park was debated in the Savoy departmental council and presented to the local population and the elected officials. And it was again in terms of a park devoted to local development that Pierre Dumas defended the law in the National Assembly. For the first time in France, elected officials defended the creation of a national park.

In Paris as well, the two projects elicited very different reactions. After a long period of procrastination, the Forestry Agency (DGEF), essentially in charge of environmental protection and the mountains, decided to seriously study the question. The agency came down heavily in favour of a national park as seen by Marcel Couturier. The people in charge of territorial planning were much more receptive to the project promoted by Gilbert André. Haunted by the book of Jean-François Gravier, *Paris and the French Desert*, published in 1947, they wanted to encourage initiatives capable of rebalancing national development and stimulating the economy in regions of encroaching abandonment. They therefore decided to launch a preliminary study on the creation of a national park in Savoy. The study was put in the hands of an architect and town planner, Denys Pradelle (1913–1999), who had played a major role in creating the Courchevel ski resort.

## Local Development not a Priority in the 1960 Law

Rather than select one or the other of the competing projects promoted by Marcel Couturier and Gilbert André, Denys Pradelle cleverly decided to combine the two. The national park was to ensure both protection and development, and thus make everyone happy. However, given the functional models prevalent at that time, it was decided to physically separate protection and development. Denys Pradelle proposed a concentric structure with three zones. First, reserve zones, small in size and completely protected, where only authorized persons, notably scientists, would be allowed to enter. Secondly, a central zone, called the 'heart' of the park, which corresponded fairly closely to the borders drawn by Marcel Couturier, well protected but where human activities would be possible. And finally, a much larger peripheral zone where a form of development consistent with the protection of the central zone would be implemented. Gilbert André was not convinced that zoning was a good idea because he had always imagined the national park as a whole, rather than as an assembly of neighbouring zones, devoted either to protection or to development. But on the whole, the work by Denys Pradelle was well received in that the different groups considered that their point of view had been taken into account and their expectations would be met. Everyone agreed that no progress could be made without a solid legal basis. The unfortunate experience with the Pelvoux National Park, created after the first world war in the Écrins and still without any legal status or resources (see Zuanon 1995), had made it clear that a law on the creation of national parks was indispensable.

The study by Denys Pradelle was presented under the auspices of both the Agriculture ministry and the territorial planning agency. However, the drafting of the law was assigned by the Prime Minister, Michel Debré (1912–1996), exclusively to the DGEF under the Agriculture ministry, that is the territorial-planning agency did not participate in preparing the law. As noted above, the Forest Agency did not share the development ambitions of Gilbert André or Denys Pradelle and perceived the national park above all as a means to protect wildlife and natural areas in general. The head of the DGEF, Yves Bétolaud (1926–2003), was in contact with many naturalists and had already drafted texts in favour of protection, notably the 1957 decree creating nature reserves. In preparing the explanatory text for the law, he did stress a number of times the originality of the French approach to national parks. Contrary to their foreign predecessors, the French national parks were not to be exclusively dedicated to environmental protection. However, the first article in the 1960 law declared that protection was the priority. The law stipulated that a peripheral zone may be set up around the central zone. What was originally an obligation had become an option. Starting with two opposing projects, the park had gone through a phase of equilibrium between the two and ended up as a law favouring one of the initial projects. The decree signed on 31 October 1961, creating the public structure for the park, confirmed the situation. No legal advantages were granted to the peripheral zone and the corresponding human and financial resources were virtually inexistent.

The difference between the projects and the actual texts initially resulted in strong reactions in the concerned areas. Elected officials in Savoy and local representatives of the central government warned Paris about the risks of the national park being rejected because it did not correspond to the project presented to the public and was much more oriented towards environmental protection. In response, the government highlighted the jobs that would be offered by the national park. This argument held weight in the beginning of the 1960s, when certain major projects such as the Tignes dam had been almost finished, and the ski resorts were just starting to be built and to hire. The government also made efforts to reassure certain population groups worried by the fact that the State had gained control over land that was essential for grazing and whose value for tourism was just becoming apparent. Negotiations were launched with the various entities involved (Électricité de France, the armed forces, the first promoters of ski resorts, town councils) to make sure that the future boundaries of the central zone did not block their projects.

Most of the actors in the process rapidly noted that the zoning and the lack of attention paid to harmonious development did in fact have certain advantages. The naturalists had always preferred strong protection over a limited area rather than limited protection over larger areas. With the national park, they obtained a truly preserved central zone, without hunting or tourist installations other than the shelters for hikers. The promoters made sure that the central zone did not cover the sectors they had targeted and the protective measures in the peripheral zone were so weak that the promoters could do as they pleased. Convinced that the peripheral zone would end up looking like the suburbs around large cities if no measures were taken, Denys Pradelle was the only person who attempted to reinforce the applicable regulations and save something of his original compromise. Gilbert André would have nothing to do with the national park and devoted himself entirely to the village of Bonneval-sur-Arc where he was elected mayor in 1956. That is where he attempted to put his ideas into application and to create, on a miniature scale, the park that he had imagined for the Alps as a whole.

The Vanoise National Park was officially created by a decree dated 6 July 1963. The central zone covered almost 53,000 hectares and the peripheral zone approximately 145,000 hectares (see Figure 7.2). The local actors most involved in its creation were very aware that it had little to do with the project initially presented to the inhabitants and that it would be much less active in local development than had first been imagined. That is why they decided to found an association, the *Friends of the Vanoise National Park*, intended to establish closer and more confident relations with the inhabitants than the public structure created to manage the park.

Though the development role of the national park, compared to the initial proposals, was significantly reduced in the 1963 decree, it should be noted that the public structure had a board comprising local elected officials who could thus weigh on park policies. The elected officials were certainly a small minority on the board, however the fact that the board included representatives of both the state and the population was novel and fairly uncommon in the highly centralized France of the 1960s.

**Park Launch and a Double Retreat**

Two experienced forestry engineers, both of mountain origin, Maurice Bardel (1912–1982) and Alfred Moulin, were appointed to manage the public structure. Because it was the first national park in France, they had no guiding examples and the texts, though they provided indications, did not contain answers to all their questions. That is why they took off, each on their own, to visit foreign parks, gather information and find ideas. From their world tour, they came back with the conviction that no existing structures could be imported to the Vanoise. In Sweden, the national parks had been set up in virtually uninhabited regions, whereas in Japan, population densities were quite high, and so on. The two men decided that the best solution was to follow their intuition and adapt to the circumstances.

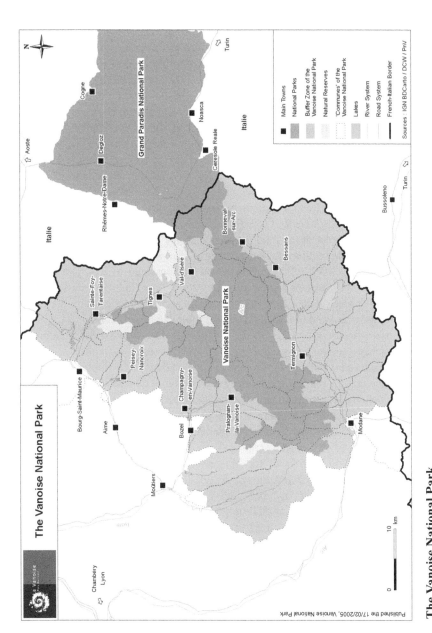

**Figure 7.2    The Vanoise National Park**
*Source: The Vanoise National Park Authority*

The park was created in July. Workers were rapidly hired to post the central zone before the start of the hunting season on 1 September. Each morning, the men started off with buckets of paint and marked the most visible rocks at fairly regular intervals with the national colours, blue, white and red. They managed to finish just in time. When the hunting season opened, the central zone was clearly identified. The first act of the park was a symbolic fence, a sign of things to come. And park officers were indeed occupied primarily with the protection of the central zone.

The bulk of the work carried out by the recently hired officers was inside the newly posted central zone. For a number of years, the human and financial resources of the park were devoted, almost entirely and quite literally, to the construction of the central zone. With the exception of Pralognan and Val d'Isère, mountaineering was not common in the Vanoise and shelters were few and far between. To house the future visitors, the park initiated the construction of shelters linked by trails that often had to be created. High trails, that the local population had never needed in the past, were created, offering panoramic views of the valleys (see Préau 1972: 144). The main purpose of the new trails was not to make the daily life of the inhabitants easier, but to show the range to hikers.[2] Protection of hooved animals was a second major occupation of the officers in the field. In particular, the surveillance and monitoring of the rare ibex, that were all located in the central zone whose borders were virtually identical to the 'ibex park' of Marcel Couturier, represented a large part of their working hours as well as their professional identity. Indeed, the ibex, chosen as the park emblem, was also on the badges worn by the officers. The latter were essentially the sons of farmers and had occupied many jobs (plumber, bus driver, factory worker, worker on hydroelectric projects, farmer, and so on) before working for the national park. Park management encouraged them to follow courses as naturalists and many became, over the years, outstanding experts in Alpine plants and wildlife. It was not a rare occurrence that the park officers helped herding farmers, who were often family members or neighbours, but such help was an individual initiative and there were no programs to support pastoral farming or local activities in general. The park officers were thus essentially occupied with welcoming the public and protecting the central zone, and almost never active in the peripheral zone. But virtually all of them were from the peripheral zone and lived there, and as a result, the park was nonetheless present in the zone. A few were very active in local affairs, for example as town councillors or even deputy mayors. But for many years, they were not allowed to be mayors. Charles Maly (1939–2001), elected mayor of Termignon in 1977, finally obtained permission even though he was employed as a park inspector in the Haute-Maurienne valley.

The scientific committee, created by the board of governors during its very first meeting, also reflected the priority given to protection over local development. It comprised a majority of scientists active in the natural and life sciences, and only a small number in the human and social sciences. The goal was to better understand the environments to be protected and efforts were made to inventory the many species. On the other hand, very little work was done on the local communities in spite of the fact that they were undergoing rapid change.

---

2   However, it is necessary to mention the *Entre-deux-Eaux* road at Termignon, that considerably shortened travel to the high summer pastures.

A little less than ten years after the park was created, the 'Vanoise affair' reinforced the retreat into the central zone and the focus on protection. A promoter, Pierre Schnebelen, wanted to create a ski resort in the central zone. He was backed by Pierre Dumas, president of the park board, and by Joseph Fontanet, a board member, who thought that the project was a means for the Haute-Maurienne valley to catch up economically. They argued in favour of the project before the board and the representatives of the various ministries received instructions to follow suit. With a one-vote majority, the board 'decided not to reject the project'. This decision was not accepted by the scientific committee and the *Friends of the Vanoise National Park*. A protest campaign was organized with conferences held in many towns, articles published in the press, and petitions signed by tens of thousands of people against the 'amputation' of the first French national park. The affair was discussed at the highest levels of government and the project finally abandoned. For the first time in France, the nature protectors (the word 'ecologist' was just starting to be used) had succeeded in organizing themselves and in contesting a decision. That constituted a major event and Florian Charvolin documented its importance, notably for the creation of the Environment ministry in January 1971 (see Charvolin 1993). However, the victory of the ecologists hardened the opposition between protection and development. The relations between promoters of tourist installations and the supporters of the national parks became much more difficult and the creation of new national parks took place in a much more conflictive atmosphere than was the case for the Vanoise park. The promoters understood that the park was more solidly installed than they had thought and they did not present any more major projects. But they did constantly try to grapple small zones along the edges of the central zone, with as a result a firm reaction on the part of the park officers who defended their borders. The protection of the central-zone borders and of the park's 'integrity' became a major issue.

In the middle of the 1980s, the 'second Vanoise affair' broke out. This time, the national electricity company requested permission to install a pumping station, comprising three dams including one in the central zone at the only point where its borders dropped down to the valley bottom. Once again, the park board initially granted its approval and, once again, the scientific committee and the *Friends of the Vanoise National Park* protested against the project and succeeded in striking it down. On two occasions, the members of the scientific committee managed to counterbalance the board of governors and the central state, and in the process the naturalists gained considerable weight. They had earned their reputation as the true defenders of the park and of its moral and physical integrity.

In the period between its creation and the second Vanoise affair, the national park struck off on a path having little to do with the original goals. The intent of the state to create, in the same place and at the same time, a national park and major ski resorts for an international public, resulted in a hardening of positions rather than the desired balance between protection and development. The park certainly proved its capacity to resist promoters, but twenty years after its creation, it was devoting most of its time and energy to protection and to protecting itself. The *Friends of the Vanoise National Park* association had difficulties in carrying out its tasks. A bulletin was published, hikes for children were organized in the park and the association initially gained a large number of members, including important people in the mountain community such as Roger Frison-Roche and Samivel. But it never succeeded in truly associating the local population in the park project. As the park gained in solidity, the association languished and lost members.[3] The attempt to counterbalance the highly protective approach of the park by an association concerned rather with local development never really produced results.

However, the park was not totally absent from the peripheral zone. It did withdraw in the face of the major ski resorts which rapidly amassed large financial resources and held the park's capacity to intervene in the economic field to be negligible. But park officials succeeded in establishing close relations with towns left out of the development efforts for mass tourism, most of which were located in the Haute-Maurienne valley where the funds allocated by the park were appreciated. A number of 'park welcome centres' (exhibitions and shelters) were set up and the park financed economic and cultural activities, such as the cultural weeks in Aussois and Pralognan, in which park officers and members of the scientific committee played an important role. The park also contributed throughout the peripheral zone to renovating religious buildings and restoring traditional roofs covered with large, flat stones.

### A Phase of Opening

The park initiators were almost completely absorbed by its construction and defense. Toward the end of the 1980s and increasingly in the 1990s, the park started to open up to the outside world and to new policies, thus drawing somewhat nearer to the original ideas of the designers. There are a number of reasons that explain this phenomenon.

First of all, protection of the central zone was essentially a success, notably concerning the protection of large wild animals. There were approximately 50 ibex in the beginning of the 1960s and over 2,000 today. The number of chamois has also increased spectacularly. A number of species that had disappeared have now come back. The lammergeyer, released in a number of spots over the Alpine arc in an international reinstallation program, spontaneously settled in the Vanoise where a number of couples are now present and reproducing. Wolves have been observed since the end of the 1990s and lynx have been noted in a number of spots. Though certain species continue to require a great deal of work and monitoring, some time could nonetheless be devoted to other activities.

At the same time that the increase and diversification of the large wild animals was noted in the central zone, the damage to natural areas outside of the strictly protected zone continued, notably in the peripheral zone where the ski resorts continued to grow and to install lifts after the Winter Olympic Games in Albertville in 1992. Park officials were of the opinion that they could and indeed had to play a greater role outside the central zone. Other protected zones had been set up and,

---

3   The *Friends of the Vanoise National Park* association was disbanded in January 2002.

similar to the Vanoise National Park, they started by first establishing themselves firmly, before looking beyond their immediate borders. Informal, then increasingly formal contacts were made between the structures, first on the national level, then the international, and finally an Alpine network of protected zones was set up in 1995. The idea emerged that it was necessary to preserve, and if need be, recreate 'ecological corridors' between protected zones. Because they had the means and felt the need, the managers of protected areas increasingly attempted to take action in areas where they had no legal grounds, that is they had to convince rather than impose. In the Vanoise, the expression 'park zone' was used starting in the 1990s to designate the central and peripheral zones taken together, the intent being to no longer separate the two as clearly as in the past. This tendency has steadily grown since then, similar to the desire to establish partnerships with local governments, including those where tourism is very prevalent, and with certain actors, notably in the agricultural sector. The managers of protected zones are not the only ones to learn, with more or less success, how to negotiate and make agreements. Other environmental sectors also tend now to round out the authoritarian attitude, for which the state has less and less means, with attempts to find innovative solutions collectively hammered out (see Aggeri 2000).

The constitution of a numerous and diversified team at park headquarters and the changes in the job of sector managers considerably helped in shifting from a park which, simply to survive, was obliged to turn in on itself, to a park desirous of opening up to the outside world and to new policies. A fairly large number of persons were hired at the end of the 1980s and during the 1990s, including agronomists, who were particularly sensitive to pastoral farming and its evolution, and positions were created specifically for communication and the peripheral zone in 1988 and 1989 respectively. A person specifically in charge of pastoral farming has been employed since 2001. In addition to their work, these men and women brought with them new perceptions of the park and a desire to establish relations with local actors. Some of them adopted the ideas of the park designers, particularly those of Denys Pradelle, to whom tribute was significantly paid in 2002. In the field, the sector managers have for a number of years devoted an increasing percentage of their time to projects with local actors.

That being said, it was not just the park that was changing, but the surrounding society as well. From the 1950s to 1970s, a handful of visionaries faced off with a local population of whom a majority was indifferent or hostile, and the protected zones resembled besieged citadels. Though threats certainly continue to weigh on the protected zones, the situation has nonetheless changed. Ecological ideas and concerns now permeate society. The actual substance of those ideas and concerns may be doubtful, but promoters and elected officials have learned to 'add a touch of green' to their speeches and even occasionally to their acts. Along with the managers of protected zones, they can now imagine establishing partnerships and going beyond the long-standing oppositions.

However, the opening up of the park is not without difficulties and not always easily accepted. Within the park, some officers fear that the primary mission of protection will suffer from greater interaction with local actors. By going beyond the central zone and attempting to compromise with development, the national park could lose its *raison d'être*. Outside the park, the inhabitants and their representatives, now accustomed to seeing the park reduced to the central zone, are worried about an extension of its influence to the peripheral zone. The reluctance on both sides contributes to slowing progress. The contracts signed with farmers number less than half a dozen and partnerships with local governments remain minor.

## A Draft Law under Constant Surveillance

In 2003, MP Jean-Pierre Giran submitted to the Prime minister a report on the national parks based on over 300 auditions of persons involved in existing or planned parks.[4] Noting a profound attachment to the national parks, he recommended that their status not be modified, i.e. they must continue to be run by the state. But he also drew attention to serious failures and severe criticisms. And observed that no national parks have been created since the Guadeloupe park in 1989, that the current projects[5] are blocked, that local elected officials often express a feeling of being dispossessed and excluded, and finally, that the peripheral zone, 'characterized by an absence of any legal standing',[6] did not contribute to true partnerships between the parks and local governments. That is why he argued in favour of a major revision of park goals and operations in a new law to replace the 1960 law, judged out of date in the current, largely decentralized France. It is necessary, in his opinion, to provide the national parks with a double goal of environmental protection and sustainable development. In spatial terms, the law must explicitly define parks as composed of two parts, a protected zone, corresponding to the current central zone, that he proposed to name the 'heart' of the park, and a second zone devoted to sustainable development. Local governments would have no choice concerning their inclusion in the 'heart', but would remain free to sign the environmental and sustainable development charter covering the rest of the park. In short, 'regulations in the heart, the contract in the peripheral zone, that should be the system' (Giran 2003: 22). The author then presented a list of 25 proposals, including more elected officials on the board of governors, the obligation to elect the board chairman from among the elected officials, and the creation of a local community committee in addition to the scientific committee to advise the board.

The Giran Report, presented to the public in June 2003 for the 40th anniversary of the Vanoise National Park, was used by the government as a blueprint in drafting the law on national and marine parks presented to the National Assembly in May 2005. It stipulates that national parks will conduct preservation and development policies for the heart and charter zones respectively. Of importance is the fact that park inhabitants with property rights in the protected zone may be dispensed with observing the regulations 'to ensure, in compliance with the protection mission assigned to the park, the normal enjoyment of their rights' (Art. L. 331-4-2).

---

4    For a presentation of the Giran report, see (Laslaz, 2005).

5    A number of national parks (Hauts de la Réunion, Guyane, Mer d'Iroise and the Calanques between Marseille and Cassis) are in the planning stage, some for a number of years.

6    Giran, 2003: 10.

A number of mountain and environmental-protection associations such as Frapna, France Nature Environnement (FNE) and Mountain wilderness opposed the law, that they considered 'influenced by economic considerations and personal interests using cultural arguments to advance their cause'.[7] The protection groups remain wary of elected officials, deemed incapable of seeing beyond short-term economic gain, and the idea of them gaining a majority on park boards has provoked great concern. The National Environment Union (SNE) launched a petition 'to save the national parks' and hopes to have the draft law withdrawn. Several thousand signatures were presented in November 2005 to MP Giran. The reactions against the law were relatively strong and the debate on development, even sustainable, in the national parks is certainly not finished.

The law was voted by the Assembly on 1 December 2005 and will have been examined by the Senate in January 2006. The government has declared the text urgent, i.e. it will not pass again before the Assembly.

**Conclusion**

MP Giran concluded the foreword of his report to the Prime minister saying, 'It is, above all, [the] alliance between protection of natural areas and sustainable development, between the 'natural' and the 'cultural', between our heritage and our future that [the national parks] must protect and perpetuate'. Though they used other terms (sustainable development, so common today, but unheard of then), Gilbert André and Denys Pradelle said exactly the same thing at the end of the 1950s. At the start of the 1960s, legislators decided to make protection the priority. Today, at the start of the new millenium, when awareness of environmental problems has never been greater, they are rebalancing the goals of national parks in favour of development. The paradox is perhaps only apparent. The history of the Vanoise National Park makes clear that the revision of the 1960 law is part of a long-standing process (geographical opening and new policies) and that, to a certain degree, it is a return to the ideas of some of the initiators. As presented in the draft law, the national parks bear a very close resemblance to the compromise between the projects of Marcel Couturier and Gilbert André, presented by Denys Pradelle. Generally speaking, the protected zones reveal a tendency noted in other environmental policies, where the state, aware of the size, prevalence and complexity of the problems to be solved with limited means, encourages actors to change their behaviour and invent new practices, without relinquishing its capacity to regulate and coerce.

It remains to be seen just what sustainable development can mean in what is, for the time being, the peripheral zone of the Vanoise National Park, which has the highest concentration of ski resorts in Europe, if not in the world. Clearly, a law recognising the need to control and limit development outside the strictly protected zones is too late for the Vanoise. It also remains to be seen if the heart of the park will remain a truly preserved zone or if the exceptions for people with property rights will open the door to multiple interpretations and uncontrollable damage. Finally, it

---

7   See sne.objectis.net/petition-pn_html.

remains to be seen what financial resources the national parks will in fact receive, at a time when the environment has suffered repeated budget cuts.

## References

Aggeri, F. (2000). Les politiques d'environnement comme politiques de l'innovation. Annales des Mines / Gérer et comprendre, 60: 31–43.

Charvolin, F. (1993). L'invention de l'Environnement en France (1960–71). Les pratiques documentaires d'agrégation à l'origine du Ministère de la protection de la nature et de l'environnement. Thèse, Grenoble, Université Pierre Mendès-France, Paris, École nationale supérieure des Mines: 503 p.

Giran J.-P., 2003. Les parcs nationaux. Une référence pour la France. Une chance pour ses territoires. Rapport au Premier ministre. 89 p.

Laslaz, L., 2004. *Vanoise. 40 ans de parc national. Bilan et perspectives.* Paris, L'Harmattan.

Laslaz, L., 2005. La réforme des Parcs Nationaux entre rapport et débats. Interrogations sur l'évolution des emblèmes de la protection de l'environnement en France. *Revue de géographie alpine*, 93/2: 111–115.

Mauz, I., 2003. *Histoire et mémoires du parc national de la Vanoise. 1921–1971 : la construction.* Grenoble, Revue de Géographie alpine. Collection Ascendances.

Mauz, I., 2005. *Histoire et mémoires du parc national de la Vanoise. Trois générations racontent.* Grenoble, Revue de géographie alpine. Collection Ascendances.

Préau, P., 1972. De la protection de la nature à l'aménagement du territoire : l'expérience caractéristique du parc national de la Vanoise. *Aménagement du territoire et développement régional.* Volume V: 119–171. Institut d'études politiques. Grenoble, Cerat.

Zuanon, J.-P., 1995. Chronique d'un parc oublié. Du parc de la Bérarde (1913) au parc national des Ecrins. Grenoble, Revue de la Géographie Alpine. Collection Ascendances.

## Internet Links

Draft law on national and marine parks: www.ecologie.gouv.fr.

Chapter 8

# Preserving the Man Made Environment: A Case Study of the Cinque Terre National Park, Italy

Stefan Kah

## The Man Made Landscape: Cinque Terre

The Cinque Terre (literally: 'Five Villages' – Monterosso, Vernazza, Corniglia, Manarola and Riomaggiore) are a 10km stretch of land along the Ligurian coast that was declared a national park in 1999. It is the cultivated landscape surrounding the villages that makes the Cinque Terre so special. The combination of a favourable climate and very difficult geographic conditions for crop growing led to the creation of about 1,400 hectares of cultivated land in the form of terraces. About 7,000km of dry-stone walls were erected here through constant collective work over the centuries. They were built without any kind of cement, creating a series of numerous terraces from sea level up to an altitude of hundreds of meters. They were primarily used for viniculture, but the terraced vineyards alternate with strips of olive trees, vegetable gardens and lemon production.

Though today the villages in the Cinque Terre give the impression of typical picturesque fishing villages, their inhabitants originally were farmers and lived in the hills. Only when the sea became safe from pirates they moved towards the coast and settled along the mouths of the rivers, starting with merely complementary fishery and using the sea mostly to trade their products, usually wine.

In the last century difficult access to this terraced land and profound structural change led to a gradual abandonment of crop growing in this area. This in turn led to the reduction of cultivated land to about 150 hectares today. The old abandoned terraces now are at a constant risk of landslides, due to the lack of necessary repair works. The reduction in cultivated land was accompanied by a decrease in population from around 8,000 to merely 4,400 people today. However, many of the buildings remain in use as second homes or summer residences.

## Touristic Boom

Parallel to the decline of the traditional economic system based on wine, olives, fruits and, to a certain degree, also fishing, coincided with the discovery of the Cinque Terre as a tourist destination. They had initially been a place for painters and poets

**Figure 8.1    The Cinque Terre National Park**
*Source : Drafted by author*

who were looking for seclusion and simplicity, then for members of the alternative movement and dropouts. It was not until the 1970s, when the landscape was already rapidly degrading, that the number of tourists increased significantly. But the Cinque Terre held on to their image as an insider tip (see Hennig 2001, 9) into the 1990s, despite the dramatic increase in the number of tourists visiting the area. However, the estimation of the number of visitors differs significantly, with some of them going up to several million a year. For the month of August 1999 the National Park Organization even gives an estimated number of 3,350,000 visitors.

In any case, the number of tourists increased at least three-fold over the last 20 years (see Richter 2001, 42), to the extent that today's mass tourism puts a strain on the local infrastructure, especially in the summer months.

Tourism in the Cinque Terre centers around hiking and experiencing the unique (cultural) landscape. There is only one decent sized sandy beach, situated at Monterosso, which is also the only village that can easily be reached by car. However, the Cinque Terre offer a lot of routes to hikers most of them with interesting contrasts between the hills and the sea. The most famous path is the *Via dell'Amore*, connecting Manarola to Riomaggiore, parts of which are cut out of the steep cliffs overlooking the sea.

So the Cinque Terre can clearly be distinquished from other tourist sites along the Italien coast that mostly focus on beach holidays. Nevertheless they have to cope with the problems created by an enormous number of tourists in a limited space. Access to the Cinque Terre has always been difficult; only a few roads, built mostly in the 1970s, connect the villages with the hinterland. It is the effective railway along the coast that provides thousands of tourists a day with access to the area. Luckily the five villages are situated along a railway line of international importance, connecting Tuscany with Genoa and the French Cote d'Azur. But this fact also allows many tourists, mostly foreigners, especially from overseas, to visit the Cinque Terre at just a day trip or to use it as a stopover. A problem is given as well by the lack of accomodation possibilities, so that many visitors stay overnight elsewhere. This results in a relatively low number of officially registered bednights of about 200,000 a year.

## A Local Initiative as a Trigger

As early as the 1970s a committed local group from the village of Riomaggiore tried to counteract the imminent degradation of the landscape. Some of their projects were later taken over by the Park Organization. Recognizing viniculture as the foundation of the Cinque Terre's traditional landscape a first phase of the project saw the implementation of measures to support this crop.

Only giving back an economic value to the cultivated land was going to make its preservation on a larger scale possible. Up to that point, wine had been produced by a large number of winegrowers, but just in small quantities, and compared to other up-and-coming wine regions it was also of poorer quality.

On the one hand the measures of the initiative refered to organizational aspects. It turned out that the future of the wine terraces depends on cooperation among the wine producers and on coordination by a supervising institution. The first step was the creation of the *Cooperativa Agricoltura Cinque Terre* (CACT) in the year 1973, a cooperative of winegrowers that made possible the creation of two regional quality wines. The Italian label *Denominazione di Origine Controllata* (DOC) then could be assigned to the classic white wine *Bianco delle Cinque Terre* and the rare precious raisin wine *Sciacchetrà*. The second step was the construction of the *Cantina Sociale* in 1983, a union of common wine cellars, where the wine could be produced in a controlled way and to specific standards (see Casavecchia and Salvatori 2002, 51).

On the other hand there were measures that concerned technical modernization aspects. To reach especially remote and steep wine terraces the CACT built small rack railways, a system originally intended for alpine environments. To compensate for occasional dry periods a simple irrigation system based on widely branched out hoses was installed. By resulting in a relatively low impact on the landscape both investigations helped to establish again a functioning viniculture.

But the cultivation conditions were still too difficult to allow the Cinque Terre viticulture to compete with success on the wine market in the future. The only chance was to concentrate on quality instead of quantity, focusing primarily on the way the grapes are cultivated. According to the motto of the cooperative 'One landscape – one wine', the consumer 'buys' a part of the landscape and apparently is willing to pay a higher price. But this specific market share would be too small to guarantee a reliable income for the Cinque Terre in the future so the local initiators recognized that wine had not the potential to be the Cinque Terre's main economic basis.

None the less in 2002 140,000 litres of wine have been produced by the CACT out of 200 tons of grapes, which represented 80% of the total production of about 300 part-time winegrowers. The wine is sold mostly directly to the consumers in the Cinque Terre thanks to an enormous demand, not only by its inhabitants but also by tourists. In any case a growing amount is sold to wine merchants in Italy and abroad, especially in the US.

Decisive on the way to a national park was the realization of the *Progetto Integrato Mediterraneo*, an integrated development project in the eastern part of the Cinque Terre. This project for the first time included touristic aspects and set an impulse for further initiatives.

Above the village of Riomaggiore about 100 *rustici*, typical structures used in agriculture and spread out widely over the terraces, had been identified as the point of departure for a project that combined tourism development and landscape preservation. These simple stone buildings can be found all over the Cinque Terre, where a lot of them were facing destruction at the time. Only some were still in use during the grape harvest period or as weekend or summer residences.

Because of relative ease with which some of these structures could be accessed, mainly thanks to a new road, nonlocal investors showed interest in purchasing them. To avoid their transformation into an interchangeable holiday resort, the community of Riomaggiore convinced the owners of the *rustici* to hand over their management to the CACT. In doing so, the land not only remained property of the Cinque Terre's inhabitants, the profits of leasing the rustici also remained in the region. This principle of adopting plots of land was later integrated into the concept of the national park.

In the pilot phase of this limited project the buildings were renovated, terraces and paths rearranged and an old pilgrims' hostel was newly opened to the public. After the economic success of the project became evident, other private initiatives started to convert old *rustici* into holiday homes. Offering accommodation in the midst of the terraces became an attractive investment and others started again to invest in rearranging their terraces (see Besio 1995, 37).

## From Regional to National Park

In the seventies ecological aspects entered the public consciousness in Italy, as they did in most European countries. Large protected areas had already existed in Italy for a long time – the first national park, Gran Paradiso, was founded in 1922. But the law 616/77 from 1977 marked an important change in Italian nature conservation policy by transferring the competencies for the protected areas from the state to the regions. Though national parks remained a national issue, the new law led to the creation of numerous new protected areas (today there are about 130 Italian regional nature parks), and to a conceptual debate about the term 'park'. This cleared the way for innovative concepts for protected areas, and eventually this also came to include the Cinque Terre National Park.

The Region Liguria also pursued the aim to establish regional natural parks and thus created an inventory of suitable areas, one of these including the Cinque Terre. At the time protection was still defined as the preservation of areas with 'environmental interest' (Rossi 2001, 253). By also recognizing the Cinque Terre as suitable for the status of protected area, the aims of protection changed: '…the presence of typical terraced areas cultivated with vine (…), represents a unique phenomenon created by human activity and is to be preserved as such' (Rossi 2001, 253).

But it was not until the 1990s that the first protected area was established, the Parco Naturale Regionale delle Cinque Terre. This park was decreed by Ligurian law and comprised a territory much larger than the actual Cinque Terre. Although it officially still exists today, this regional park was never accepted by the communities affected. The interests within the park were too different; some of the touristic resorts were afraid of restrictions for their economic development, especially regarding building projects. The Cinque Terre communities also refused to accept the park. They saw in it an improper use of the brand Cinque Terre as an instrument of touristic marketing for a large area, that did not correspond with the brand. The challenges that faced some of the participating touristic resorts were in fact significantly different from those in the Cinque Terre.

In 1997, the Cinque Terre municipalities established their own 'bottom-up' park concept and pitched it against the proposal of the Liguria region. The new initiative suggested a smaller and more homogenous territory which was supported by the UNESCO's decision to recognize the Cinque Terre as part of 'Mankind's World Heritage'. Also in 1997, the *Riserva Marina delle Cinque Terre* (Cinque Terre Protected Marine Area) was established, covering exactly the strip of coast in front of the five villages and already defining a possible inclusion in a future national park in its constitution.

Surprisingly, there was almost no opposition from inside the project region to the park, only the local hunters and fishermen expressed criticism. More controversial was the discussion about the extension of the park. The fear of future restrictions on the one hand and the hope for benefits for tourism thanks to the label national park on the other hand created some conflicts, but in the end a compromise was reached. Eventually, five communities agreed to take part in the national park: the three actual Cinque Terre municipalities (in terms of administration, Manarola belongs to Riomaggiore, and Corniglia to Vernazza) with their complete territory and the surrounding towns of Levanto and La Spezia with small coastal areas.

The Parco Nazionale delle Cinque Terre was established by decree of the 15th of April 1999 and inaugurated in December of the same year. With 4.226 hectares it is the smallest of at the moment 20 Italian national parks and the first one that became at the same time part of the UNESCO World Heritage (see Rossi 2001, 258).

**The Park Organization as Development Agency**

The five villages of the Cinque Terre have a common identity and today they face similar challenges, though they have different backgrounds historically. But none the less, throughout history, this relatively homogeneous region never had its own competencies or any leeway regarding decision making, being stuck between the official administrative units of province and community. Only through the establishment of the national park did the Cinque Terre become a region in the sense of own competencies.

The Park Organization takes advantage of the new opportunities to act jointly together and sees itself as more than just the administration unit of a protected area. As a development agency it started to bring on various projects regarding all aspects of the economic and social life in the Cinque Terre, going far beyond simple preservational aspects. Thanks to a law from 1991 (*Legge quadro sulle aree protette*) encompassing protected areas, it is possible for national parks to create their own income and use it at their own risk (see Maglia and Santoloci 2000, 662). This possibility is crucial for most of the park's projects.

While the administrative aspects are covered by the Park Organization, including tourist marketing and preservation issues, other work, such as running the tourist offices and the national park shops, is assigned to so called *cooperative*. These are small enterprises favoured by the tax system, designed to help especially younger people entering the job market.

The Park Organization had the advantage of having at its disposal a local population that was mostly in favour of the park and its objectives. There was no need to explain the need of acting to save the Cinque Terre cultural heritage. The population was also well aware of the endogenous potentials they had and demonstrated a strong relationship with their territory. In addition to this the Cinque Terre can rely on a long tradition of collective and voluntary work, from medieval times through to the work of the CACT in the seventies.

So the activities of the Park Organization are based on various successful projects established since the *Progetto Integrato Mediterraneo*. The three biggest challenges were the preservation of the landscape, the (often too) dynamic development of tourism and the strain on the infrastructure, especially concerning the volume of traffic.

The success of the already running projects was never doubted, but the long-term preservation of the Cinque Terre could only be achieved by interconnecting the various activities. The Park Organization created a coordinated development programme based on the principles of sustainability and initiated a local Agenda 21 process. Its focal point was creating a sustainable form of tourism by using its positive effects and avoiding possible negative outcomes.

These aims should be achieved by: 1. defining sustainability for this specific case, 2. developing a catalogue of suitable measures and 3. continuous supervision of this process.

The resulting definition of sustainable tourism for the Cinque Terre was:

> to develop a result-focusedd plan that guarantees the long-term profitability of a tourist destination and its compatibility with ecological, socio-cultural and economic needs (translation by author). (Parco nazionale delle Cinque Terre)

Unfortunately, the benefits of introducing sustainability into processes can often only be noticed after many years. This led to a catalogue of measures that contained not only long-term challenges. Small projects easy to realize were also developed to make the process more comprehensible, as for example by implementing the use of energy-saving bulbs in the Park's offices and in hotels. This visibility created acceptance within the population and thus won time and confidence for the realization of long-term projects.

The three projects described in the following are crucial to the strategy of the Park and they are examples of projects with enormous effects on the ecological, social and economic situation.

## Preservation of the Cultural Landscape

PROSIT is a project that forms part of the European Union's programme LIFE (*L' Instrument Financier pour l'Environnement*). Its aim is the preservation of the characteristic wine terraces by making the visitors participate directly in the maintenance of the terraced landscape. Thanks to a study by the University of Genoa a map of the complete terraced territory could be provided, serving as a decision-making basis for the selection of plots of land suitable for rearrangement.

One of PROSIT's elements, *Recupero Terre Incolte* (literally: 'Recovery of unused land'), aims to entrust those who actively participate with fallow vineyards for a period of 20 years. The idea for this project relates to activities already run by the CACT before the foundation of the park.

On the basis of a regional law from 1978 the owners of agricultural land are obligated to cultivate their terraces or to hand their utilization rights over to the Park Organization. In this kind of adoption the owner does not get paid a rent, but receives his land back after 20 years in a renovated state. The adopter is obligated to rearrange the terraces, to plant grapes and to produce wine respecting strict ecological guidelines.

Because of the labour-intensive work only small plots of land (up to 3,000 m²) can be adopted, but along with these the adopters get the right to use the *rustici* on the land. To keep a certain standard and to respect local building traditions the renovation of these *rustici* – including the installation of a self-sufficient energy supply system based on sun or wind – is carried out by the Park Organization, who ask the adopters for reimbursement of the costs (approximately 17,500 €).

In most cases the duties of winegrowing are assigned to locals for payment, thus creating extra income for the Cinque Terre. This initiative led to a remarkable demand among visitors. Up until now about 4,500 requests have been made to the Park Organization, so that it has encountered difficulties in finding enough persons able and willing to carry out the repair works.

Another initiative of PROSIT is the establishment of work camps on the terraces for interested people, mostly groups of students. These are called *Università del paesaggio* (landscape universities) and offer free board and lodging in exchange for work on the terraces, combined with lectures on the value of landscape and ecological aspects.

**Environmental Quality Certification**

The *Marchio di qualità ambientale* ('Environmental Quality Brand') is a voluntary certification instrument for tourist sector enterprises in the Cinque Terre National Park. It is adressed to three categories of businesses: a. hotels, b. rooms and apartments for rent and c. restaurants.

The aim is to create conditions that allow the enterprises not only to offer a service of high quality but at the same time to guarantee the observation of certain environmental criteria. These are among others: waste recycling, use of ecological detergents, offering organic breakfast and – in the case of restaurants – offering a typical menu accompanied by local wine.

In exchange for respecting these criteria the Park Organization offers organizational support, especially admission to the Park's marketing instruments like the website and the tourist infopoints. Some obligatory objectives of the brand must be implemented during the first year, enhancement objectives have to be implemented gradually. The brand is adressed especially to small enterprises; by constant coaching by the Park Organization the participants build up a network of enterprises. In this way the Cinque Terre can compensate in part for the lack of big hotels, offering for example co-ordinated marketing management for the majority of the tourist businesses.

By introducing a brand adressed primarily to private rooms for rent, the Park Organization also pursues another aim. The project supports an accommodation offer that, compared to hotels often owned by non-residents, generates a relatively high income effect among the local population. The acceptance of the brand is also high, as already from the beginning of the project about 80% of all enterprises decided to participate.

**Sustainable Mobility**

As already mentioned the accessibility of the Cinque Terre National Park is difficult and practically only by train. In the course of establishing the park a sustainable traffic concept had to be found. The existing railway represented already an ecological, efficient and successful mean of transport, whose potential was not yet fully used. Nevertheless, today about 80–85% of the growing number of visitors use public transport and also a lot of the inhabitants renounce the use of their car most of the time. The railway became the basis of the concept, which includes a type of admission ticket to the Cinque Terre's services and generates income for the Park Organization. The *Carta Cinque Terre* (Cinque Terre Card) is a combination ticket for public transport into and within the national park, also including admission to some of the most important hiking trails, the Nature Observatory Centre, an internet access point and several other services.

In this way the Cinque Terre Card has become a form of – however, still non obligatory – admission ticket to the National Park. If tourists want to use the hiking trails, even if only the famous but short Via dell'Amore, they have to buy a ticket. As it also includes the train journeys, the decision about the means of transport to reach the Cinque Terre is easily made.

The Card is available for one (5,40 €), three (13,00 €) or seven days (20,60€) and is sold inside the Cinque Terre as well as at the surrounding train stations like in La Spezia, where most of the visitors arrive by long-distance trains to start their journey into the national park. Thanks to the new income the offer in public transport could be expanded. Several buses with engines based on electricity or methane gas now connect the villages at sea level with small settlements up the hills. In this way a system of internal mobility could be established, offered free of charge to the inhabitants of the Cinque Terre. The price of the ticket is relatively low, considering that just the journey by train itself would cost almost the same amount. Obviously the most important project partner is the Italian Railway Association, who initially planned to close the ticket offices in four of the five villages. Now the stations have been transformed into visitor centres and the ticket offices are run by the Park Organization. In exchange for this, the Railway Association forgoes a part of the income generated by the Cinque Terre Card, making the low price possible.

An important effect of this project is that it generates income for the Park Organization. The profit of the Cinque Terre Card can be invested into the preservation of the cultural landscape and the improvement of tourist services. By this it helps to reduce the dependence on subsidies from outside the park.

### The National Park as an Integrated System

As demonstrated with the three main projects, tourism is used as an instrument to preserve the cultural heritage of the Cinque Terre National Park. It is the impulse for regional development and is the necessary exogenous input which activates the endogenous potentials. As shown in Figure 8.2 the relationship between tourism and the Park is defined by numerous interconnections.

Within this system the impact of tourism triggers processes that go far beyond merely economic aspects. They obviously generate added value, but thanks to several instruments made available by the coordinating Park Organization, a part of it is used to follow specific aims (e.g. preservation, improvement of living conditions). Thus a functioning local economy contributes to maintaining the unique selling proposition of the tourist destination Cinque Terre, the cultural landscape.

The result is an interdependence between tourism and cultural landscape, or rather the Cinque Terre as a whole. The national park with its attractive cultural landscape represents the tourist attraction, and tourism in turn contributes to maintain the landscape. Finally, in the Cinque Terre National Park tourism does not endanger the local cultural heritage, but supports its preservation.

### The Philosophy of the National Park

For a long time, protected areas were seen as measures to preserve a certain territory from human impact, but the argumentation has now shifted. More and more parks are founded to preserve a certain form of human impact, for example a certain kind of cultural landscape. The protected area itself offers the institutional framework and

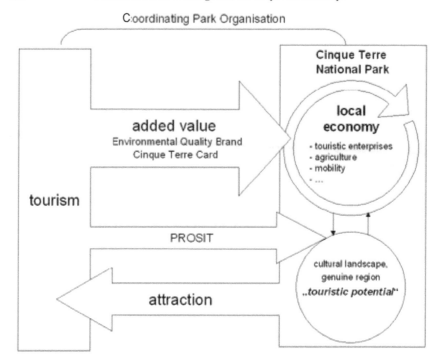

**Figure 8.2    Integrated system, Cinque Terre National Park**
*Source: Drafted by author*

provides a tourist brand, and with its instruments and subsidies becomes the basis for social and economic development in the future.

In Italy the Cinque Terre National Park is the first park created to safeguard a landscape that is mostly man-made. In certain aspects it corresponds more to the concept of a biosphere reserve or various kinds of natural or regional parks, as they can be found especially in Central Europe. In the Cinque Terre National Park the focus is on the man-made environment, shown also by the park's motto: *Il parco dell'uomo* (The man-made park). Consequently, experiences from other protected areas are of little use.

In the Cinque Terre the issues differ from those of other protected areas. Agriculture is often an antagonist of park projects, but in this case it is not only one of the most important partners but even one of the explicit aims.

Further, the challenge of the Cinque Terre National Park is a different one. It is not the stagnation or decline, as it is often the case in other peripheral areas where parks are used as a measure of development policy. It is the high dynamics, on the one hand concerning the rapid degradation of the landscape, and on the other hand the touristic boom, which brings with it the risk of more disadvantages than advantages, if not instrumentalized as illustrated above.

The Cinque Terre National Park is obviously a special case, but some elements of its development strategy could also be applied to other protected areas. The preservation of cultural landscape by introducing models of adaptation or certification for small tourist businesses could indeed be transferred into other contexts. But those areas need to dispose of their own specific potentials. If, instead, the motivation for protected areas is only the lack of alternatives in development, without attractive potentials, simple copying of ideas will probably fail.

## References

Besio, M. (1995), 'Recupero del territorio delle Cinque Terre', *Parametro*, 1995:211, 37–57.

Bonati, L. (2001), *Il mare segreto delle Cinque Terre*, (Sarzana: Grafiche Lunensi).

Casavecchi, A. and Salvatori, E. (2002), *Storia di un paesaggio*, (Riomaggiore: Cinque Terre National Park Press).

Hennig, C. (2001), *Cinque Terre und ligurische Küste* (Badenweiler: Oase).

Kah, S. (2003), 'Tourismus als Katalysator integrierter Regionalentwicklung – das Beispiel Cinque Terre', *Arbeitsmaterialien zur Raumordnung und Raumplanung*, 03:222, (Bayreuth: University of Bayreuth Press).

Kah, S. (2005), 'Cinque Terre – Tourismus als Katalysator integrierter Regionalentwicklung', *Standort*, 29:2, 71–5.

Maglia, S. and Santoloci, M. (2000), *Il codice dell'ambiente* (Piacenza: Casa Editrice La Tribuna), 662–79.

Parco Nazionale delle Cinque Terre (undated), *Parco Nazionale Cinque Terre. Presentation of the Park*. Unpublished.

Richter, M. and Block, M. (2001), 'Vielfalt in den Cinque Terre (Ligurien)', *Geographische Rundschau*, 53:4, 40–47.

Rolla, S., De Franchi, R. and Pasini, L. (2003), 'La salvaguardia e la rivitalizzazione del sito culturale nelle Cinque Terre di Liguria', *local land & soil news*, 03:7/8, 13–14.

Rossi, L. (2001), 'Il parco delle Cinque Terre. Dibattito istituzionale e sociale', *Rivista Geografica Italiana*, 108, 247–65.

Saretzki, A., Wilken, M. and Wöhler, K. (2002), 'Lernende Tourismusregionen', *Tourismus-Beiträge zu Wissenschaft und Praxis*, 3 (Münster: LIT Verlag), 1–79.

Schemmerer, A. (2004), *Der Nationalpark Cinque Terre (Italien). Wechselbeziehungen zwischen Nationalparkanspruch und touristischer Wahrnehmung*. Unveröffentlichte Zulassungsarbeit an der Universitaet Erlangen-Nuernberg. (Erlangen).

## Internet Links

Italian National Parks: http://www.parks.it
National Park Cinque Terre: http://www.parconazionale5terre.it

Chapter 9

# National Parks and Rural Development in Spain

Andreas Voth

## Introduction

In the last three decades of democratization in Spain, a significant consolidation and regionalization of environmental policy took place and is reflected by an amazing proliferation of protected areas of different categories, which become incorporated into the proposed Spanish Natura 2000 Network occupying more than 20% of the country's territory. The localization of most areas with some kind of protection in disfavoured rural districts, especially in the mountains, requires a progressive transition from traditional policies of passive conservation to new approaches of active conservation, integrating the local population and surrounding territories and converting the protected areas into instruments of sustainable development (see Troitiño 2005). A recent shift of paradigm can be observed in protection policies in Spain, especially in nature parks and biosphere reserves, towards new instruments of environmental planning and management taking into account the close interrelations between protected natural areas and their social and economic environment. Even the national parks are no longer seen as areas isolated from the rest of the territory, and their peripheral protection zone is increasingly regarded as part of a surrounding area of socioeconomic influence offering opportunities for the promotion of sustainable development. The challenge to overcome the traditional confrontation between conservation and development and to demonstrate their compatibility also has to be seen against a background of a profound structural change in rural areas under the influence of reforming European agricultural and environmental policies. This article deals with the national parks in Spain, presenting an overview of their historical development, legal framework, present-day problems and their increasing relation with rural development initiatives.

## Historical Background: From Covadonga to the Present National Park Network

In the European context, Spain was one of the first countries to create national parks. In this survey only some crucial aspects of this long historical evolution are mentioned, but for more details the exhaustive documentation presented by Fernández and Pradas (2000) is recommended. External and internal factors determined the beginning of conservation policy in Spain, where the diffusion of the idea of Yellowstone was received by a changing policy and society (see Solé and Bretón 1986). The history of national parks in Spain started relatively early with the direct introduction of the American model, but with only few contacts with similar initiatives in other European countries, where social and political sensibility to nature conservation had already spread and to a greater extent (see Casado 2004). In Spain, the roots of nature conservation are closely linked to traditional forest policy. Another important factor was the new estimation of high mountain landscapes and nature due to the beginning of Pyreneeism and the proliferation of hiking associations. In Catalonia, early proposals to protect the Mountains of Montserrat (1902) as a national park were not successful. A famous protagonist and pioneer of nature protection in Spain was Pedro Pidal, Marquis de Villaviciosa (Asturias), mountaineer and senator with great political influence, who visited Yellowstone and Yosemite in 1915. His personal knowledge about conservation models applied to the first American national parks had a direct impact on the conversation policy in Spain. In his philosophy of national parks, Pedro Pidal presented the natural heritage as a paradise necessary to protect and stated:[1] 'If we do not guard the possessed paradise between the lost paradise and the promised paradise, we do not deserve, like Adam, to have any paradise.' The protected natural landscape is compared frequently to art monuments and also contains certain notions of patriotism and religion. Right from the start, national parks have been the subject of controversial discussions. Pidal claimed the responsibility of the State to create national parks:[2] 'Instead of national parks of Catalonia, which due to belonging to a region cannot be named "national", we have to create national parks of Spain. Belonging to Spain they will belong also to Catalonia and Andalusia, to Galicia and Murcia.'

Spain participated early in the declaration of protected areas, publishing the first Spanish National Park Act in 1916 to preserve some selected areas from human exploitation. The law defined national parks as 'sites or places of the national territory exceptionally picturesque, wooded or rough', which are protected by the State, respected and made accessible to visitors.[3] Two years later, two national parks were declared: Covadonga in the Cantabrian Mountains, and Ordesa in the Pyrenees. The name of the first Park is a reference to the historical site of Covadonga in Asturias, and its inauguration by the Spanish King Alfonso XIII in 1918 had to coincide with the twelve hundredth anniversary of the legendary Battle of Covadonga, where the Christian Resistance defeated the Arab invaders in the rough terrain of the Picos de Europa – a historical event and place considered the starting point of the Spanish Reconquista. The idea of the national park was associated with different values, as the words of Alfonso XIII show:[4] 'we are going to do something unique in the World: to unite the art of Nature with Religion and History, at the birth place of the Nation … This is Covadonga: Spain.'

Central goals of park declarations were nature protection as well as promotion of tourism that was not perceived yet as a potential aggression to nature. Nevertheless,

---

1    Pedro Pidal 1917, quoted in Gómez Mendoza (1992).
2    Pedro Pedal, 1916, quoted in Fernández and Pradas (2000, 75).
3    Ley de Parques Nacionales, 7th of December 1916.
4    Quoted in Menéndez de la Hoz (1999, 4).

the founders of the first national parks soon realized that 'Covadonga was not Yellowstone, and Ordesa not Yosemite' (Casado 2004, 25). The main problems resulted from the resistance of local inhabitants to cease their traditional land use rights and from the limited capacities of the Park Administration. No further national parks were declared up to the 1950s, whereas the smaller and more tolerant Natural Sites received increasing attention and were considered to be less conflictive and more appropriate to the Spanish conditions than national parks of American style (see Fernández and Pradas 2000, 163). In the 1920s, several areas were named 'Natural Site of National Interest', a protection category especially favoured by the scientist Eduardo Hernández-Pacheco, another key protagonist of the early conservation history in Spain complementing but also competing with the political initiative of Pedro Pidal. Forestry engineers played an important role in the early evolution of nature conservation and in the elaboration of a long list of Natural Sites to be proposed for protection (see Mata Olmo 1992).

The effects of the Spanish Civil War interrupted the evolution of the conservation policy, but in 1954 the first two volcanic landscapes on the Canary Islands[5] were declared national parks, indicating their growing interest for tourism. A year later, another high mountain area was protected in the Pyrenees of Catalonia, the Aigüestortes y Lago de San Mauricio National Park, which is known for its characteristic glacial modelled landscape of granite rocks and numerous mountain lakes, and which is quite different from the limestone mountain massif of Ordesa and the Picos de Europa with their strong Atlantic influence. All early national parks represent high mountain landscapes estimated for their symbolic, grandiose, spectacular, mystic or magic character. The Mountains Act of 1957 substituted the National Park Act of 1916 and attributed the management of national parks to the forest administration.[6] To the end of the decade, the increasing international interest in ecosystem research and protection drew the attention on the remaining wetlands, too long ignored by conservation policy and seriously threatened by expanding drainage and irrigated agriculture. First initiatives to protect part of the Guadalquivir river marshlands in Andalusia were directly related to the foundation of the World Wildlife Fund (WWF) in 1961. Thanks to international pressure and financing, a growing area of these wetlands of greatest ecological value could be protected, leading to the declaration of almost 35,000 hectares of Doñana as a national park in 1969. Doñana has become the most famous Spanish national park on the international level, but also the most conflictive and difficult to manage. Nature conservation was not compatible with the dominating economic development policy of that time. The parallel but uncoordinated promotion of the national park, intensive agriculture, and mass tourism in neighbouring or even overlapping areas caused serious problems still not fully resolved. Polemic discussion about human pressure on wetlands also encouraged the declaration of the small Tablas de Daimiel National Park in the upper Guadiana river basin on the Meseta plains of La Mancha in 1973.

---

5   Due to their specific characteristics and evolution, the four national parks existing on the Canary Islands are neglected in this article, although they are included in the Network of National Parks of Spain.

6   Ley de Montes, 8th of June 1957; Decreto 485/1962, 22nd of February 1962.

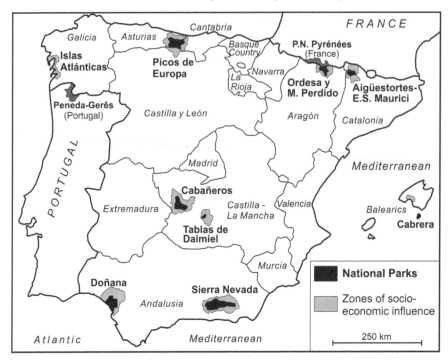

**Figure 9.1     National parks of the Iberian Peninsula**
*Source: Drafted by author after Troitiño et al. 2005, modified; Europarc-Spain; MMA*

After a long period of stagnation, one of the last laws[7] approved by the old dictatorial regime in 1975 revived the debate on protected areas in Spain, although it was a law of transition in a changing political context and substituted by the basic legislation[8] of 1989. The beginning processes of democratization, social and economic change and the political status of Autonomous Communities granted to the regions gave a fresh impetus to the amplification and declaration of national parks and especially to the diffusion of other categories of protected areas. The transfer of responsibilities for environmental legislation to the regional level caused a real boom of protected areas and also a change of their objectives and tasks. The nature park became the preferred category of protected areas, and priorities shifted gradually from landscape and ecosystem conservation to regional development. This trend offered new development perspectives to peripheral rural areas with severe structural problems. In national parks, however, aspects of ecological representativeness and diversity gained in importance.

Following the international trends and responding to a growing public awareness and social demand for nature conservation, the State tried to complete the National Park Network including some representative ecosystems still missing. The first

---

7     Ley 15/1975, 2nd of May 1975, Espacios Naturales Protegidos.

8     Ley 4/1989, 27th of March 1989, Conservación de los Espacios Naturales y de la Flora y Fauna Silvestres.

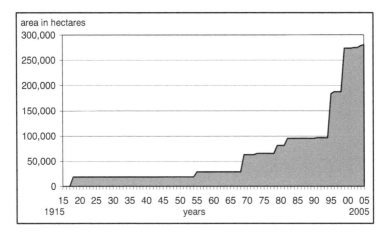

**Figure 9.2    The growth of the area declared as national parks in Spain (1915–2005)**

*Source: Drafted by author*

national park enclosing territories of three adjacent Autonomous Communities, named Picos de Europa, resulted from the extension of the emblematic Park of Covadonga in 1995. In 1999 an even larger area was declared national park in the Sierra Nevada, the highest Mediterranean mountain range that already had been protected before as a nature park by the regional administration of Andalusia. Another important piece of the Network is the Mediterranean woodland in the continental centre of the Iberian Peninsula, protected in the National Park of Cabañeros founded in 1995. This landscape with scattered trees and abundant wildlife is known as 'the Spanish Serengeti'. Also coastal and maritime ecosystems were included into the National Park Network: the small Mediterranean Island of Cabrera in 1991 and some Atlantic Islands of Galicia in 2002. The crossing of ecological (Cabrera) and administrative borders (Picos de Europa) indicates new trends in the conservation policy, as well as the coordination between Spanish protected areas and the cross-border park Peneda-Gerês, the only Portuguese national park.

The spatial distribution of the Spanish national parks (see Figure 9.1) represents the history of the country's conservation policy, from the initial attraction of high mountain landscapes up to the recent protection of wetlands, Mediterranean woodlands and maritime areas (see Cabero 2001). Park amplifications and the declaration of larger parks led to a remarkable growth of the total area protected as national parks in Spain in the last decade (se Figure 9.2), and this process has not yet stopped definitely, as the ongoing discussion about new national parks shows, especially about the Sierra de Guadarrama, a mountain range North of Madrid, and the Mediterranean oak woodlands of Extremadura in the area of the present Monfragüe Nature Park. Like the Sierra Nevada, also the Sierra de Guadarrama was proposed for declaration as a national park as early as in 1930, when the protection and management of such large areas still were considered to be too difficult (see Mata Olmo 1992, 1077). At present, thirteen parks are integrated in the National Park Network, including the four national parks of the Canary Islands which are especially

**Table 9.1    National parks in Spain, including Canary Islands**

| National Park | Declaration | Park Area | Municipalities of National Parks | | Visitors in 2004 |
| --- | --- | --- | --- | --- | --- |
| | (Year) | (Ha) | (Nr.) | Area (Ha)   Population | (Nr.) |
| Picos de Europa | 1918 | 64,660 | 10 | 124,637 | 15,060 | 2,221,761 |
| Ordesa - Monte Perdido | 1918 | 15,608 | 6 | 85,820 | 1,974 | 582,800 |
| Aigüestortes - Maurici | 1955 | 14,119 | 10 | 141,130 | 11,336 | 341,759 |
| Sierra Nevada | 1999 | 86,208 | 44 | 266,526 | 63,785 | 558,489 |
| Doñana | 1969 | 54,252 | 4 | 143,470 | 25,618 | 391,536 |
| Tablas de Daimiel | 1973 | 1,928 | 3 | 81,340 | 30,289 | 128,630 |
| Cabañeros | 1995 | 40,856 | 6 | 183,850 | 6,089 | 70,740 |
| Archipiélago Cabrera | 1991 | 1,318 | 1 | 21,033 | 367,277 | 73,540 |
| Islas Atlánticas | 2002 | 1,195 | 5 | 26,835 | 365,966 | 182,394 |
| Teide (Tenerife) | 1954 | 18,990 | 14 | 116,820 | 264,244 | 3,540,195 |
| C. Taburiente (Palma) | 1954 | 4,690 | 9 | 56,772 | 48,717 | 367,938 |
| Timanfaya (Lanzarote) | 1974 | 5,107 | 2 | 34,635 | 12,858 | 1,815,186 |
| Garajonay (Gomera) | 1981 | 3,986 | 6 | 37,180 | 19,580 | 859,860 |
| *All 13 National Parks:* | *-* | *312,917* | *120* | *1,320,048* | *1,232,793* | *11,134,828* |

*Sources: Organismo Autónomo Parques Nacionales; García and Asensio 2004; Troitiño et al. 2005; own compilation*

exposed to the pressure of tourism. More than 1.2 million people live in the areas of socioeconomic influence, in the 120 municipalities with part of their territory in one of the national parks that receive about 11 million park visitors per year (see Table 9.1). The complex relationship between national parks, local population and visitors requires an adequate legal framework for planning and management.

## National Parks in the Legal Framework of Protected Areas in Spain

The Law 15/1975 for Protected Natural Areas diversified the protection formulas and brought a new dynamic into the declaration of protected areas. The democratization process and the transfer of environmental legislation responsibility to the autonomous regions called for a general legal framework in Spain, given by the Law 4/1989 for the Conservation of Natural Areas and the Flora and Fauna. This basic law established four types of protected areas – the Park (not specified), the Nature Reserve, the Natural Monument and the Protected Landscape – but it left much room for interpretation to the autonomous regions and favoured a heterogeneous evolution of protected areas in Spain. The regional administrations developed their own laws and created their own specific protection categories. This uncoordinated procedure has led to a great diversity of denominations and planning instruments for protected areas in Spain. Some authors record up to forty different types of protected areas (see Troitiño 2005, 234). Presenting a review of the protection formulas in Spain, also Florido and Lozano criticize the excessive number of different terms and definitions that make it difficult to compare parks declared on the regional level (see Florido and Lozano 2005).

The result is an unconnected conjunction of regional networks with heterogeneous typology, approaches and objectives (see Mulero 2002). The category 'national park' is neglected by regional legislation and less used than other types of protected areas that do not suppose such strict limitations to human activities as national parks.

According to Law 4/1989 (Art. 13), parks are 'natural spaces little transformed by human exploitation or occupation that, in view of their beautiful landscapes, represent ecosystems or unique flora, fauna or geomorphologic formations, possess ecological, aesthetic, educative and scientific values whose conservation deserves preferential attention'. National parks are defined as 'natural spaces of great ecological and cultural value' whose conservation is of general interest for the Nation and that represent the natural heritage.[9] A Directive Plan (Plan Director) was developed to maintain the image and coherence of the growing Network of National Parks and to define the general performance criteria.[10] This document describes the network as 'an integrated system of protection and management of the selected best examples of the Spanish natural heritage. It is shaped by the integrated national parks, the normative framework, the material and human resources, the institutions and the system of relations necessary for functioning'.[11] A public sector body assigned to the Ministry of the Environment is the Autonomous National Parks Organization (OAPN) that has developed and coordinated the planning and management of the Network of National Parks since 1995. The network's purpose is not only the conservation of the national parks, but also 'the interchange of knowledge and experiences in sustainable development'.

The Plan Director is a basic planning framework including the following objectives:[12]

- To favour the consolidation of the Network of National Parks, its internal coherence and a homogeneous and coordinated planning and management.
- To contribute to the system of nature protection and conservation in Spain, incorporating the national parks into national and international conservation strategies.
- To establish the necessary guidelines for conservation, public use, investigation, training, education, increasing social awareness and sustainable development.
- To favour the development of public consciousness and appreciation of national parks and to channel social participation in the process of decision making.
- To define and develop a framework for cooperation and collaboration with other national and international administrations.
- To promote the image and the external projection of the Network of National Parks.

---

9  Ley 4/1989, Art. 22.1.

10  Real Decreto 1803/1999, 26th of November 1999, Plan Director de la Red de Parques Nacionales.

11  Real Decreto 1803/1999, Chapter 2.1.

12  Real Decreto 1803/1999, Chapter 1.

In addition, the Plan Director defines the criteria for areas to be protected as national parks:[13]

- Representativeness: a national park must be highly representative of the natural system to which it belongs.
- Extension: parks must include an adequate extension of territory to allow its natural evolution and to maintain its functionality and ecological processes.
- State of conservation: predominating conditions of naturalness and ecological functionality with little human intervention.
- Territorial continuity, without fragmenting elements.
- Settlements: must not be included; only in case of justifiable exceptions.
- Buffer-zone protection: parks must be surrounded by a territory that can be declared peripheral zone of protection.

The Directive Plan also indicates the two relevant instruments of planning and management to be elaborated and applied individually to each national park of the network, which are the Plan for the Regulation of Natural Resources (PORN) and the Steering Plan for Use and Management (PRUG). The PORN has to define the state of conservation of the natural resources of the protected area in order to establish the levels of protection needed and the policy criteria for socioeconomic activities compatible with nature conservation. On this basis and according to the Directive Plan, the PRUG is developed as the main planning instrument, which must include the following contents:[14]

- Norms and general directives for use and planning of the park.
- The zoning concept that limits areas of different use and establishes norms of application.
- Determination and programming of activities for conservation, investigation and diffusion.
- Economic estimation of investment in infrastructure, conservation, investigation and public use.
- Identification of activities compatible with the aims of the national park.

Nevertheless, the elaboration of these planning documents often suffers difficulties and delays, due to political disputes, frequent changes of the legal frameworks and a lack of acceptance. Some authors criticize the Directive Plan as well as the specific PORN and PRUG planning documents for their still dominating vision from inside the parks lacking an integrative view of the functional relations between parks and surrounding areas (see Troitiño 2005, 258). Other critics reject the Plan Director for keeping up the traditional American model of national parks which excludes the presence of human activities and settlements and makes the creation of new national parks difficult in the European context (see Campos and Carrera 2005, 28). The relatively high conservation standards and the centralized administration of the

---

13  Real Decreto 1803/1999, Chapter 2.2.
14  Ley 4/1989, Art. 19.4.

**Table 9.2     Protected areas in the autonomous regions of Spain**

| Region | Protected Areas (km²) | Protected Area (%) | Nr. of Protected Areas* | Nr. of National Parks* | Area of National Parks (km²) | National Parks: area as % of all protected areas |
|---|---|---|---|---|---|---|
| Andalusia | 16,053 | 18.4 | 130 | 2 | 1405 | 8.8 |
| Catalonia | 8,721 | 27.2 | 243 | 1 | 141 | 1.6 |
| Castilla-León* | 5,774 | 6.1 | 21 | 1 | 247 | 4.3 |
| Canary Islands | 3,195 | 42.9 | 146 | 4 | 328 | 10.3 |
| Castilla-Mancha | 2,137 | 2.7 | 50 | 2 | 428 | 20.0 |
| Asturias* | 1,857 | 17.5 | 53 | 1 | 245 | 13.2 |
| Valencia | 1,153 | 5.0 | 23 | 0 | 0 | 0.0 |
| Aragón | 1,110 | 2.3 | 8 | 1 | 156 | 14.1 |
| Madrid | 1,042 | 13.0 | 10 | 0 | 0 | 0.0 |
| Basque Country | 806 | 11.1 | 38 | 0 | 0 | 0.0 |
| Navarra | 747 | 7.2 | 105 | 0 | 0 | 0.0 |
| Galicia | 695 | 2.3 | 21 | 1 | 12 | 1.7 |
| Murcia | 680 | 6.0 | 19 | 0 | 0 | 0.0 |
| Cantabria* | 573 | 10.8 | 7 | 1 | 154 | 26.9 |
| Extremadura | 448 | 1.1 | 20 | 0 | 0 | 0.0 |
| Balearics | 275 | 5.5 | 14 | 1 | 13 | 4.7 |
| La Rioja | 241 | 4.8 | 2 | 0 | 0 | 0.0 |
| Total | 45,507 | 9.0 | 908* | 13* | 3,129 | 6.9 |

\* *The National Park of Picos de Europa is shared by three regions.*
*Sources: Troitiño et al. 2005; Europarc-Spain*

national parks in Spain impede an uncomplicated elaboration and application of the planning instruments.

The national parks initially were conceived as elements of cohesion in the pretended Spanish network of protected areas, but the autonomous regions did not accept them to be managed exclusively by the central administration of the state (see Mulero 2002, 66). Catalonia established a precedent obtaining an early responsibility transfer to the regional level. The problems of the Aigüestortes National Park in a region considered by separatist movements to be a Nation and the continuous responsibility conflicts about park management demonstrate the difficulties to maintain a nationwide Network of National Parks against the disintegrating pressure of regionalism. Some regional governments directed constitutional complaints against the legislation about national parks, and in 1995, a sentence of the Constitutional Court[15] forced the central and regional administrations to adopt a shared management. Modifying the Law 4/1989 about protected areas, the Law 41/1997 established a new form of joint management and financing and created the instrument of the Directive Plan for National Parks to harmonize aspects of planning and management within the network.[16] Nevertheless, some autonomous regions still continued considering the legislation about national

---

15  Sentencia 102/1995 del Tribunal Constitucional, 26th of June 1995, about the Ley 4/1989.

16  Ley 41/1997, 5th of November 1997, modifying the Ley 4/1989.

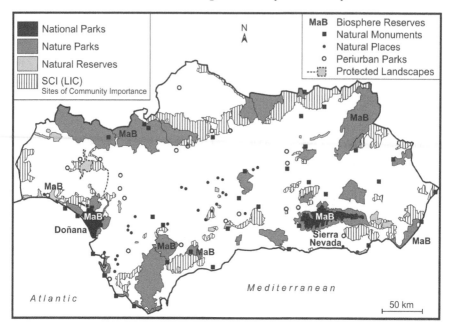

**Figure 9.3    The system of protected areas of Andalusia (RENPA)**
*Source: Drafted by author after RENPA, Consejería de Medio Ambiente, Sevilla*

parks as unconstitutional and reclaimed absolute decentralization. Finally they reached their goal,[17] so that the legislation framework of national parks has to be changed once more, including part of the Directive Plan.[18]

The regionalization process has not only led to an institutional change of the Network of National Parks, but also to a parallel evolution of regional networks and innovative types of protected areas characterized by an increasing integration of development functions. The total number of protected areas in Spain has risen up to more than 900, and national parks occupy less than 7 per cent of the total area with some kind of protection (see Table 9.2). Although many autonomous regions have established their own network of protected areas, this concept is still criticized to have a more administrative character than a territorial or functional one. Even small regions like Asturias designed a complex network of different protected area categories in addition to a national park and including the distinction of several parks as biosphere reserves (see Voth 2004a, 43). The network is an essential element of the region's tourism development plan. The case of Andalusia shows the enormous expansion of the protected areas based on the own legal framework of 1989 and the continuous efforts to complete and improve the regional network (see Figure 9.3). The national parks of Doñana and Sierra Nevada are surrounded by nature parks and incorporated into the Network of Protected Areas of Andalusia (RENPA). As the

---

17  Sentencia 194/2004 del Tribunal Constitucional, 10th of November 2004.
18  Sentencia 101/2005 del Tribunal Constitucional, 20th of April 2005, about the Real Decreto 1803/1999.

following chapter shows, the integration of national parks on the regional and on the international level has gained in importance.

## The Spanish National Parks in the International Context

The selection of representative examples of natural areas with low population density and high conservation and management standards facilitate the recognition of the Spanish national parks as IUCN Management Category II, defined as protected areas managed mainly for ecosystem protection and recreation. Several international agreements concerning nature conservation have great inpacts on national parks in Spain. Labels of international recognition constitute a significant increase of the park's image and prestige and are much sought-after to improve the possibilities of using the park as an instrument of regional development. Six national parks were recognized as Biosphere Reserves under the MAB Programme of the UNESCO, and three as Natural World Heritage Site. Sierra Nevada was already inscribed as biosphere reserve before being declared nature park and national park, while the pioneer Park of Picos de Europa obtained this title much later in 2003. Moreover, three national parks (Doñana, Ordesa and Teide) received the European Diploma of the European Council for natural areas with excellent management, and two (Doñana and Tablas de Daimiel) were accepted as internationally important wetlands under the Ramsar Convention. Doñana is the only national park possessing the entire collection of titles: after its important enlargement of 1978, the park was recognized as Biosphere Reserve in 1981, Wetland of International Importance in 1982, Protected Area with European Diploma since 1985, Special Protection Area for Birds in 1988 and UNESCO World Heritage Site in 1994. The confusing diversity of protected areas in Spain becomes even more complex by the superposition of different other coexisting protection categories of national and international rank (Florido and Lozano 2005, 75).

Of course, all Spanish national parks also belong to the proposed NATURA 2000 network of the European Union. The establishment of the NATURA 2000 network in Spain caused a drastic amplification of the territory affected by one of the different protection formula, representing almost 22 per cent of the countries land. Although Spain developed an ambitious programme of farmland reforestation and proposed a vast NATURA 2000 network, Europarc-Spain (2002) states that in general terms, the country has made very limited use of European funds available for conservation measures. LIFE-Nature funds are mainly aimed at environmental restoration and the recovery of endangered species. EU-programmes frequently used in protected areas and their areas of socioeconomic influence are also INTERREG, EQUAL and especially LEADER, a rural development initiative with great success and acceptance in the Spanish regions. Another initiative similar to LEADER and derived from its experiences is PRODER, the 'Operating Programme for Development and Economic Diversification of Rural Zones', running parallel to LEADER in Spain. Almost all rural areas are covered by projects of the LEADER or PRODER initiative that are of supreme importance especially for nature parks.

The marginal role national parks play in the local action groups reflects the weakness of relations between the protected area and the territory in which they are embedded. Nevertheless, there are also interesting cases of innovative application of LEADER around national parks, as the experiences of Tablas de Daimiel demonstrate, where the municipalities and local actors integrated in the Association 'Tierra y Agua' ('Land and Water') put into practice several projects of local development related to their common resource and problem: the water scarcity around their national park which depends on the same aquifer. The water crisis is not only explained in the park's visitor centre, but also in a special exposition and documentation centre about water set up in the neighbouring village. The Association edited an informative guidebook for visitors including the natural and cultural heritage of the municipalities in the LEADER area as well as information about the severe water conflicts around the national park (see Asociación Tierra y Agua 1998). Another example is the nearby Cabañeros National Park, where a network of new museums and services was organized to attract more visitors to the area, and a local action group uses the name of the park in a great variety of rural development projects. The information transfer between different action groups has gained in importance. In Spain, 65 municipalities with territory inside a national park founded an association[19] as an instrument for cooperation and exchange of experiences with the aim to favour participation in policies of sustainable development.

## Land Use Conflicts and Concepts of Regional Planning in National Park Areas

The national parks in Spain already experienced many conflicts but also offer new opportunities for rural development. The traditional planning instrument of zoning is used to organize the park area depending on its carrying capacity and the characteristics of the natural resources and to make land use compatible with conservation. The Directive Plan of the Network of National Parks establishes the following zoning concept with areas from higher to lower levels of protection:[20]

- Reservation Zone: the maximum level of protection and without any public use.
- Zone of restricted use: areas with a high degree of naturalness, in a good state of conservation or regeneration and able to support a certain level of public use on indicated trails.
- Zone of moderate use: predominantly natural areas with greater visitor carrying capacity and some traditional agricultural practices.
- Zone of special use: limited areas for buildings and facilities needed within the park for public use, management and administration.
- Zone of traditional settlement: created only in cases where settlements with their adjacent areas of cultivation and services exist within a national park boundary.

Zoning concepts do not help to avoid all conflicts caused by the great number of land use interests and factors of influence inside and outside the parks. The youngest

---

19  Asociación de Municipios con Territorio en Parques Nacionales (AMUPARNA).
20  Real Decreto 1803/1999, Chapter 4.3.

national park of Spain, the Atlantic Islands of Galicia, was heavily threatened and has achieved tragic fame only a few months after its declaration in 2002, when the 'Prestige' tanker accident caused an oil pollution affecting the entire shore. The evolution of Picos de Europa in Asturias, the oldest Park, is a history of problems right from the start. A first period of conflicts was mainly caused by limitations to traditional forms of land use and persisting high mountain mining up to the 1970s. While mining companies still received permissions to continue their activities inside the national park, the local population was deprived of their traditional rights (see Fronchoso 1999). In a second period of conflicts the pressure of tourism increased. Strong economic interests to create new infrastructure for mass tourism, as well as conflicts between shepherds and the protected wolfs turned the local population against the park. The relatively small group of local shepherds receives considerable amounts of subventions and compensation for a traditional and endangered activity, but without giving up their opposition against the Park (see Casas 2004), although they are better off than traditional users in any other comparable mountain areas. In 1995, the Park enlargement included territories of the neighbouring regions of Cantabria and Castilla-León and also some inhabited villages inside the national park, exactly when the conflictive struggle for the park management responsibility in Spain started. It will be difficult to improve the never consolidated management of Picos de Europa, due to insufficient cooperation between three autonomous administrations. Frequent changes of political power on the competing national, regional and municipal levels and the abuse of the national park in political disputes paralyzed initiatives like LEADER and led to a loss of development opportunities (see Rodríguez and Menéndez 2005). An Integrated Development Plan was elaborated for the wider area of socioeconomic influence in the East of Asturias on a solid scientific and empirical basis, but it was not implemented.

Casas (1999) distinguishes between parks with different levels of conflict, from parks without any type of confrontation (Cabrera and parks of the Canary Islands), up to the very conflictive parks of Doñana and Tablas de Daimiel, where the very park declaration had been a consequence of already existing land use conflicts. In the region of La Mancha, the drainage of wetlands provoked an increasing international pressure for their conversation, but the declaration of the Tablas de Daimiel National Park could not stop the expansion of irrigated agriculture, illegal wells and the pending water crisis of the area. The survival of the national park's ecosystem depends on irregular water transfers from the Tajo-Segura-Aqueduct constructed to bring irrigation water to the dry regions of Southeast Spain. In the area around the national park, farmers receive financial compensation for deciding deliberately to reduce their groundwater consumption, but these short term solutions must be substituted by a Plan of Sustainable Development (see Carrasco 2003). The visible deterioration of the wetlands reminds people not to lose the park, increasingly considered as their own resource. Local actors are changing their attitude and become proud of their park (see Casas 1999, 273).

On the basis of the 'Action Plan for Europe's Protected Areas' published by the IUCN (1994), Europarc-Spain (2002, 20) designed a complex own Action Plan to provide orientations and recommendations on different planning and management processes in the protected areas, and to favour citizens' awareness and participation concerning conservation and development decisions. In this document, protected areas are regarded as basic instruments for territorial planning and as diffusion centres of new sustainable development strategies. The Spanish Action Plan assesses that the switch towards the planning of protected areas integrated into a wider territorial framework is still very limited and more conceptual than real (see Europarc-Spain 2002, 36), so that a specific document about the present difficulties and opportunities to integrate protected natural areas in regional planning has been elaborated recently (see Europarc-Spain 2005). Environmental and regional planning legislation are responsibilities of the autonomous regions, but generally without really integrating the policy of protected areas into regional planning. Nevertheless, their planning instruments usually perceive the revalorization of the region's natural heritage as an opportunity for a balanced socioeconomic development and regional cohesion. For some protected areas, the sub-regional planning level has gained greater importance. This is the case in the area of socioeconomic influence of the Doñana National Park, where a specific document for regional planning (see Junta de Andalucía 2004) was designed to adjust a previous document to changing legislation in Andalusia and to take up the recommendations elaborated by an International Commission of Experts concerning land and water use conflicts and development opportunities. This new legal framework for planning and sustainable development of the territory is based on a detailed characterization and analysis of present uses, infrastructure and demands in the area around the national park and contains ambitious objectives and action lines to make nature conservation and different economic uses and opportunities compatible. 'Particularly, in economically depressed zones, protected areas can mean an engine generating new sources of income and thus contributing to social development and the conservation of the natural and cultural heritage' (Europarc-Spain 2002, 72).

### Changing Acceptance, Tourism and New Concepts of Sustainable Development

For sustainable development processes in national park areas, a high level of local acceptance is an indispensable condition. In spite of early prohibitions of traditional land use, the National Park of Ordesa was accepted and brought new income opportunities able to stop depopulation processes affecting most valleys of the Spanish Pyrenees (see Casas 1999, 267). The declaration of the National Park of Cabañeros was a demand of the local population and gained great acceptance. The new Atlantic Islands National Park is also widely accepted, because traditional fishing is recognized as a compatible activity and interests of the coastal population are not affected by the Park declaration. In more conflictive parks, even substantial compensations could not guarantee acceptance, and traditional concepts based on the conciliatory effects of subsidies turned out to be inadequate as a long-term solution. Some authors denounce not only the lack of information and participation of locals in planning processes, but also the 'culture of assistance' becoming firmly fixed in the population living around protected areas (Troitiño 2005, 262). In the Doñana area, a compensation policy attempted to overcome the long lasting conflicts between nature conservation and economic interests, and people continuously expected the arrival of new 'development plans' to generate a new economic upturn, as Ojeda (1993) demonstrated in his detailed analysis. Decades of uncoordinated sector policies strengthened the deep rooted idea of local communities to have a legitimate claim to compensation for renouncing a complete transformation of the area for intensive agriculture and tourism development. New concepts were needed.

The considerable growth of purchasing power, mobility, environmental sensitivity and leisure time experienced in Spain during the last two decades, as well as the need for recreation of the urban population looking for alternative destinations to the overcrowded Mediterranean coast, are factors that explain the increase of tourism in national parks and other protected areas. Political support to environmental education and tourism development in protected areas has gained in importance. In 1996, the protected areas in Spain already received at least thirty million visitors, a third part of them in the national parks (see Gómez-Limón and Múgica 2000). In spite of the questionable data reliability, visitor statistics indicate a growing number of visitors to national parks. Picos de Europa and the parks of the Canary Islands are the most visited parks (see Table 9.1). The rapidly increasing number and quality of publications about national parks reflects the growing interest in visiting them. Some visitor guides give a complete overview of the Spanish Network of National Parks (see e.g. Alamany and Vicens 2003), while most books offer detailed information about a single park. Some scientific studies give evidence of the increasing number but also of the concentration of park visits in space and time, as the case of Doñana demonstrates (see Litago et al. 2003), and other parks suffer an even stronger visitor concentration. For some points like the Lakes of Covadonga in the Picos de Europa or the Valley of Ordesa, access plans with restrictions were elaborated. The primary objective of public use concepts for national parks originally was offering facilities of environmental interpretation and education, but the increasing demand for other recreational activities and additional services requires a change in public use planning and gives private enterprises new opportunities. The number of visitor centres and information points has increased drastically in recent years, most of them managed by the park administration, but also through service concessions. All national parks have at least one visitor centre and some information points. The perspectives of economic development based on rural tourism depend on the location, accessibility, infrastructure and attractiveness of the national parks.

## The Case of Doñana

The Doñana area is certainly one of the best studied and documented territories in Spain and also the most intervened and conflictive. The protection of large areas as a national park surrounded by a nature park has saved an important part of Doñana from being transformed for intensive agriculture, mass tourism, urbanization and traffic infrastructure. Traditional activities and an increasing number of modern land use interests put the Park under pressure (see Figure 9.4). Severe land use conflicts characterized the relationship between the Park and local population and led to an unsustainable situation requiring new development concepts and institutions (see Voth 2004b). The pressure of conflicts triggered the search for innovative solutions.

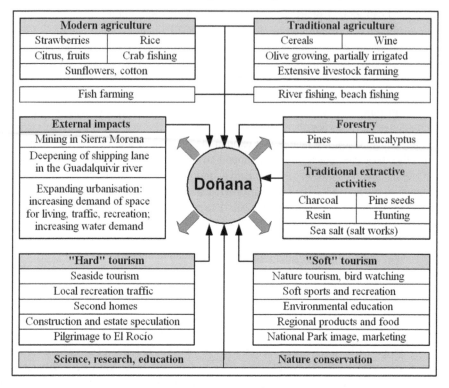

| Modern agriculture | | Traditional agriculture | |
|---|---|---|---|
| Strawberries | Rice | Cereals | Wine |
| Citrus, fruits | Crab fishing | Olive growing, partially irrigated | |
| Sunflowers, cotton | | Extensive livestock farming | |

| Fish farming | River fishing, beach fishing |
|---|---|

| External impacts | Forestry | |
|---|---|---|
| Mining in Sierra Morena | Pines | Eucalyptus |
| Deepening of shipping lane in the Guadalquivir river | | |
| Expanding urbanisation: increasing demand of space for living, traffic, recreation; increasing water demand | Traditional extractive activities | |

**Donaña**

| Traditional extractive activities | |
|---|---|
| Charcoal | Pine seeds |
| Resin | Hunting |
| Sea salt (salt works) | |

| "Hard" tourism | "Soft" tourism |
|---|---|
| Seaside tourism | Nature tourism, bird watching |
| Local recreation traffic | Soft sports and recreation |
| Second homes | Environmental education |
| Construction and estate speculation | Regional products and food |
| Pilgrimage to El Rocío | National Park image, marketing |

| Science, research, education | Nature conservation |
|---|---|

**Figure 9.4    Land use conflicts in the Doñana area**
*Source: Voth 2004, 43*

People have recognized the positive impact of Doñana on different activities around the park within a concept of sustainable development.

The judgement of the International Commission of Experts[21] tried 'to make compatible the justifiable aspirations of the inhabitants in municipalities surrounding Doñana for a better life with an integral conservation of Doñana's ecosystems, from a perspective of sustainable development'. The experts were convinced that the conservation of the park could represent a comparative advantage for the area regarding the expanding market of nature tourism and the market segments for agricultural products with a quality label or a denomination of origin. To achieve this goal, they also recommended a general improvement of infrastructure in the area and the realization of an ambitious programme of education and professional training. The start of a Plan for Sustainable Development (PDS) in 1993 was the logical consequence of the analysis of problems and opportunities. With the financial support of the EU, the regional administration implemented the plan and created the Foundation 'Doñana 21' to promote social participation and public and private actions for the sustainable development of Doñana and its area of socioeconomic influence. In the function of a new development agency for the 14 municipalities constituting the area of Doñana, the Foundation supports coordination and cooperation among

---

21  Comisión Internacional de Expertos sobre el Entorno de Doñana (1992, 4).

different entities and social agents and stimulates the participation and own initiatives of the local population. After the successful realization and evaluation process, a second PDS for the Doñana area is being elaborated in 2006. The PDS must integrate all territorial and sector policies related with the park area. The participation of local actors and rural development groups in the elaboration process and management of the plan is of fundamental importance. The experiences with the implementation of the PDS in the area of Doñana encouraged the elaboration of PDS documents also for other protected areas in Andalusia. Ten nature parks of the region received a PDS up to 2005, including the Sierra Nevada National Park.

One of the innovative initiatives carried out in the framework of the PDS is the quality label 'Etiqueta Doñana 21' created by the Foundation in 1998 to enhance the differential values that the companies of the Doñana area offer. The distinctive label is granted only to local companies that incorporate and combine systems of quality and environmental management and fulfil detailed requisites for certification. This sign of environmental respect and prestige is expected to improve the companies' competitiveness and market access and the external image of Doñana. Enterprises that like to obtain the label have to adopt ISO regulations and to demonstrate a continuous improvement of their environmental behaviour in a process evaluated by external certification. Up to 2005, seventeen enterprises and institutions already received the certification, and forty others have started the evaluation process to obtain it. Most of these enterprises are dedicated to agricultural production and marketing (35 per cent), services related to tourism (32 per cent) and to handcraft and industrial activities (12 per cent). The name of 'Doñana' also is presented as a quality label on different international tourist fairs and used for destination marketing. Another interesting approach in Andalusia is the introduction of the denomination of 'Nature Parks of Andalusia' as a trademark created by the regional administration in 2000 with the support of the EU-initiative ADAPT, following the concept of the French regional parks. Firms located in the nature parks of the region may obtain a licence to use the label for the promotion of their natural products, handicrafts or tourist services fulfilling determinate standards. The geographical name and positive image of the parks serve as an instrument of rural development and regional marketing. Tourism in protected areas plays an important role in the diffusion of the image and in the increase of sales. The National Park of Doñana receives more than 400,000 visitors annually, although less than 20 per cent of them participate in paid tours inside the park. Analysis of visitor statistics show a significant seasonality and spatial concentration of visitor flows on the major visitor centres (see Litago et al. 2003). Doñana already shows a saturation of visitor infrastructure and a great variety of other tourist attractions and facilities (see Figure 9.5).

In a workshop organized by the Foundation 'Doñana 21' (1998), local experts and the mayors of all municipalities within the socioeconomic influence area analyzed the perspectives of ecotourism and defined a model of sustainable tourism for the Doñana area. The participants approved a 'Doñana Local Charter for Sustainable Tourism' considering that 'the present trend in tourist demand, in accordance with the notion of sustainable development, requires a product which is integrated into the particular characteristics of the area'. On the international level, the 'European Charter of Sustainable Tourism in Protected Areas', approved by the Europarc

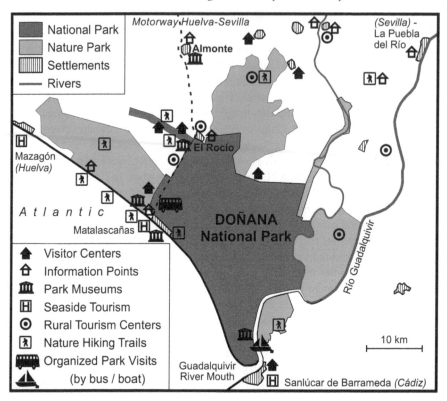

**Figure 9.5    Park tourism in the Doñana area**
*Source: Drafted by author*

Federation in 1999, consists in a successful drafting of long term strategies of sustainable tourism development with close participation of local agents, also in the socioeconomic influence area of national parks. The Nature Park of the Volcanic Area of La Garrotxa in Catalonia was one of the first protected areas in Europe participating in the introduction of the European Charter, and Sierra Nevada is the first Spanish national park obtaining this certification in 2004 as an instrument aiming at compatibility between nature conservation and economic development. The Nature Park and the National Park of Doñana jointly decided to obtain the European Charter and reinforced their coordination of administration and management structures in a broader Doñana region that surpasses the protected areas (see Pérez and Quiros 2006). The territory is identified with its natural resources and its people who increasingly participate in activities of park management.

## References

Alamany, O. and Vicens, E. (2003), *Parques Nacionales de España. 26 itinerarios para descubrirlos y conocerlos* (Barcelona: Lynx Edicions).
Asociación Tierra y Agua (ed., 1998), *Tablas de Daimiel y su Entorno* (Daimiel).

Cabero Dieguez, V. (2001), 'Espacios naturales protegidos y conservación del medio', in A. Gil Olcina and J. Gómez Mendoza (coords.), *Geografía de España* (Barcelona: Editorial Ariel), 207–221.

Campos, P. and Carrera, M. (2005) 'Conservar naturaleza y desarrollo local. Plan Director de Parques Nacionales', in *El Mundo*, 13th of February 2005, 28.

Carrasco Redondo, M. (2003), 'Daimiel como paradigma de los conflictos entre comunidades de usuarios de aguas subterráneas y conservación de humedales de alto valor ecológico', in C. Coleto, Martínez Cortina, L. and Ramón Llamas, M. (eds), *Conflictos entre el desarrollo de las aguas subterráneas y la conservación de los humedales – La cuenca alta del Guadiana* (Madrid: Mundi-Prensa), 233–254.

Casado de Otaola, S. (2004), 'Historia de los parques nacionales españoles', in V. García and B. Asensio (coords.): *La Red de Parques Nacionales de España* (Madrid: Canseco Editores, OAPN), 20–31.

Casas Grande, J. (1999), 'Conservación y espacios protegidos: el papel de los agentes locales', in M. Corbera (ed.), *Cambios en los espacios rurales cantábricos tras la integración de España en la UE* (Santander: Universidad de Cantabria), 251–296.

Casas Grande, J. (2004), 'Implicaciones socioeconómicas de la declaración y gestión de los parques nacionales en España', in V. García and B. Asensio (coords.), *La Red de Parques Nacionales de España* (Madrid: Canseco Editores, OAPN), 482–497.

Comisión Internacional de Expertos sobre el Entorno de Doñana (1992), *Dictamen sobre estratégias para el desarrollo socioeconómico sostenible del Entorno de Doñana* (Sevilla: Junta de Andalucía).

Europarc-Spain (ed., 2002), *Action Plan for the Protected Natural Areas of the Spanish State* (Madrid: Fundación F. González Bernáldez).

Europarc-Spain (ed., 2005), *Integración de los espacios naturales protegidos en la ordenación del territorio* (Madrid: Fundación F. González Bernáldez).

Fernández, J. and Pradas, R. (2000), *Historia de los Parques Nacionales españoles, vol. 1* (Madrid: OAPN).

Florido, G. and Lozano, P.J. (2005), 'Las figuras de protección de los espacios naturales en las comunidades autónomas españolas', *Boletín de la A.G.E.* 40, 57–81.

Fronchoso Sánchez, M. (1999), 'Los parques nacionales: protección y conflictos sociales en los Picos de Europa', in M. Corbera (ed.), *Cambios en los espacios rurales cantábricos tras la integración de España en la UE* (Santander: Universidad de Cantabria), 297–312.

Fundación Doñana 21 (ed., 1998), *Foundations for a Model of Sustainable Tourism in Doñana* (El Rocío, Huelva).

García, V. u. Asensio, B. (coords., 2004), *La Red de Parques Nacionales de España* (Madrid: Canseco Editores, OAPN).

Gómez Mendoza, J. (1992), 'Los orígenes de la política de protección de la naturaleza en España - La iniciativa forestal en la declaración y en la gestión de los Parques', in V. Cabero Diéguez, J.M. Llorente Pinto, and J.I. Plaza Gutiérrez (eds), *El medio rural español, vol. 2* (Salamanca: Universidad), 1039–1051.

Gómez-Limón, J., de Lucio, J.V. and Múgica, M. (2000), *Los espacios naturales protegidos del Estado español en el umbral del siglo XXI - De la declaración a la gestión activa* (Madrid: Europarc-Spain, Fundación F. González Bernáldez).

IUCN (1994), *Parks for life - Action Plan for Protected areas in Europe* (Gland).

Junta de Andalucía (ed., 2004), *Plan de Ordenación del Territorio del Ambito de Doñana* (Sevilla: Consejería de Obras Públicas y Transporte).

Litago, J., Moral, M. and Marquiegui, I. (2003), 'Evolución del número de visitas al Parque Nacional de Doñana', in L. Júdez Asensio et al. (coords.), *Valoración del uso recreativo del Parque Nacional de Doñana* (Madrid: CSIC), 17–30.

Mata Olmo, R. (1992), 'Los orígenes de la política de espacios naturales protegidos en España', in V. Cabero et al. (eds), *El medio rural español, vol. 2* (Salamanca: Universidad), 1067–1077.

Menéndez de la Hoz, M. (coord., 1999), *Guía de visita del Parque Nacional de los Picos de Europa* (Madrid: OAPN).

Mulero Mendigorri, A. (2002), *La protección de espacios naturales en España - Antecedentes, contrastes territoriales, conflictos, perspectivas* (Madrid: Mundi-Prensa).

Ojeda Rivera, J.F. (1993), *Doñana: Esperando a Godot* (Sevilla: Universidad, IDR).

Pérez, J. and Quirós, F. (2006), 'Doñana: Participación y gestión', *Medio Ambiente n°53*, Consejería de Medio Ambiente, Sevilla, 16–21.

Rodríguez, F. and Menéndez, R. (2005), *Geografía de Asturias - La reconstrucción territorial de una región de tradición industrial* (Barcelona: Editorial Ariel).

Solé, J. and Bretón, V. (1986), 'El paraíso poseído. La política española de Parques Naturales (1880-1935)', *Geo Crítica* 63, (Barcelona: Universidad).

Troitiño Vinuesa, M.A., de Marcos, F.J., García, M., del Río, M.I., Carpio, J., de la Calle, M. and Abad, L.D. (2005), 'Los espacios protegidos en España: significación e incidencia socioterritorial', *Boletín de la A.G.E.* 39, 227–265.

Voth, A. (2004a), 'Asturien - Wirtschaftskrise und neue Entwicklungsansätze in Nordspanien', *Geographische Rundschau* 56:5, 38–45.

Voth, A. (2004b) 'Nationalpark Doñana - Nutzungskonflikte und Ansätze nachhaltiger Entwicklung', *Geographie heute* 25:225, 42–47.

**Legislation Documents**

Ley de Parques Nacionales, 7th of December 1916.

Ley de Montes, 8th of June 1957; Decreto 485/1962, of 22nd of February 1962.

Ley 15/1975, 2nd of May 1975, de Espacios Naturales Protegidos.

Ley 4/1989, 27th of March 1989, Conservación de los Espacios Naturales y de la Flora y Fauna Silvestres.

Ley 41/1997, 5th of November 1997, modifying the Ley 4/1989.

Sentencia 102/1995 del Tribunal Constitucional, 26th of June 1995, about Ley 4/1989.

Real Decreto 1803/1999, 26th of November 1999, Plan Director de la Red de Parques Nacionales.

Sentencia 194/2004 del Tribunal Constitucional, 10th of November 2004.

Sentencia 101/2005 del Tribunal Constitucional, 20th of April 2005, about the Real Decreto 1803/1999.

# Chapter 10

# Market-driven Governance of Biodiversity: An Analysis of the Müritz National Park Region (Germany) from a Marketing Perspective

Markus Leibenath

## Riding the Tiger, or: Enhancing the Effectiveness of National Parks through Marketing

The loss of biodiversity and the threats to protected areas certainly can be traced back to a broad range of causes. There is no doubt that economic activities which are related to markets represent a major reason for conflicts with nature conservation. Farming, logging, tourism, transport, urban sprawl, or the extraction of mineral resources are only some of the more prominent examples of economic activities which are often made responsible for damages to nature and protected areas. Having this in mind, the idea of market-driven governance of biodiversity or of enhancing the effectiveness of protected areas through marketing may seem paradox.

Due to the common good character of nature conservation and the problem of external costs it is hard to imagine that conservation will ever be fully reconciled with the market economy. However, there are at least three reasons to overcome the traditional antagonism and to contemplate possible synergies between the two spheres.

The first and most fundamental consideration refers to the legitimacy of conservation. According to Scharpf (2004, 3), legitimacy is a functional prerequisite of forms of governance which are at the same time efficient and liberal. Legitimating arguments 'establish a moral duty to obey [...] collectively binding decisions even if they conflict with individual preferences' (Scharpf 1998, 2). There are different dimensions of legitimacy, one of which is output-oriented legitimacy. It centres on the effectiveness and efficiency of governance activities with regard to the achievement of issue-specific goals, but also on the capability of fulfilling more general functions which go beyond issue-specific problem-solving. Both the democratic system as a whole and any democratic decision are measured according to the degree they solve prevalent problems. The output-oriented dimension of democratic legitimation is linked to the postulation that collectively binding decisions should serve the common interest of the constituency (see Getimis and Heinelt 2004, 10; Scharpf 1998, 2;

Wolf 2002, 17). The need to legitimize conservation policies usually is relatively high as they represent distributive conflicts: Relatively few local stakeholders are asked to do without something, to comply with strict rules, to accept restrictions for land-use and so on for the sake of public welfare (see Möller 1995, 29). Seen from a local perspective, the legitimacy of a conservation policy will rise if it can be demonstrated that it benefits not only the national population's interests, but serves also local needs of income and jobs instead of being a mere obstacle to economic development (see Leibenath 2001, 26-8; Leibenath *forthcoming*).

The second consideration is linked to the overall political climate in which environmental policies are embedded. More and more conservationists realize how difficult it is to get the required political support for their ambitions solely on the ground of ethic argumentations and are striving to demonstrate that conservation can also be economically profitable. The political agenda in Europe is largely dominated by liberal economic thinking and by the overall goals of economic growth and creation of jobs. This is for instance reflected by the so-called Lisbon strategy of the European Union which aims at making the EU by 2010 'the most competitive and dynamic knowledge-based economy in the world, capable of sustainable economic growth with more and better jobs and greater social cohesion' (European Union 2000). These are the framework conditions which nature conservationists have to cope with. At the same time it is an opportunity to get away from asking for moral support and to show instead what benefits nature conservation has to offer society both economically and in terms of quality of life. Thus, a market-oriented approach to conservation can play an important role in legitimizing existing protected areas in a broader sense and in finding the most efficient approaches to the conservation of biodiversity (see Leibenath and Badura 2005, 4).

The third consideration is rooted in the ongoing processes of Europeanization and Globalization. The accomplishment of the European internal market, the enlargement of the EU and the abolishment of trade barriers worldwide in combination with novel information and communication technologies increase interregional competition. This fosters market-oriented approaches to regional development. What is more, globalization darkens the economic perspectives of peripheral regions as their structural disadvantages become even more apparent. While the pros and cons of globalization can be disputed at an overall political level, it is a matter of fact for an individual region. Thus, there is no other choice for actors within a certain region than to confront themselves with the increased competition. At this point, marketing comes into play. To develop and implement marketing strategies for entire regions already is as indispensable as on other contested markets, for example those for consumer goods (see Leibenath 2001, 47-9; Manschwetus 1995, 6, 19pp.). Protected areas can play an important role for the marketing of a region as they may improve the region's attractiveness for potential tourists, investors, and new citizens.

In sum, attempts towards integrating protected areas and market-oriented regional development can improve both the legitimacy of conservation at different levels and the competitive edge especially of peripheral, rural regions which don't dispose of many other assets than unspoiled nature or landscapes. In turn, higher degrees of legitimacy and acceptance of conservation policies can raise the identification of the local population with conservation goals and reduce violations of protection regulations. This is how marketing can enhance the effectiveness of protected areas like for example national parks. On the other hand, an accelerated economic development of protected area regions poses new threats from an ecological point of view. Therefore it is a constant challenge to keep economic and conservation concerns in balance.

The intention of this article is to show what experiences with regional development have been made in the Müritz National Park Region in the late 1990s and to analyze the outcomes from a marketing perspective.

As a basis, some general information on national parks in Germany and on their potential for the marketing of regions will be given in the next paragraph. In the following the case study will be presented with major attention to specific strategies and measures that have been implemented in the study area to make the national park a catalyst of regional economic development. Finally, some general conclusions on the marketing of national park regions will be drawn.

*National Park Regions – a Product like Any Other?*

*National Parks in Germany*   More than 130 years ago, the Yellowstone National Park was established as the first national park worldwide in 1872. The term 'national park' was used from the outset although it was only legally fixed in 1899. The designation of the Yellowstone National Park was inspired by three ideas:

- To preserve a pristine landscape of national importance
- To protect the area against any human alteration by the country's highest-ranking conservation authority, and
- To allow for public access to the area under certain preconditions (Jungius 1985: 10).

However, originally the territory of the Yellowstone National Park was not perceived as a natural but rather as a cultural monument by the first US-American expedition in 1869 (see Runte 1987, 35). Consequently, the national park idea evolved to meet cultural needs. The initial impetus behind preservation was the search for a 'distinct national identity' (Runte 1987).

There are fifteen national parks in Germany today (see Table 10.1) which have been established in three phases. The first German national park was set up in 1970 in the Bavarian forest at the German-Czech border. Three other parks followed until 1989. The second phase covered only the pivotal year 1990 when five national parks were designated at once in the frame of the so-called GDR[1] National Park Programme. This legislative initiative was launched in the turbulent last months of the GDR during the political turnaround of 1989/90. Eventually five more parks were established in a third phase after 1990.

In Germany, nature conservation falls principally into the competence of the *Länder*. This means that there are in practice significant differences in the

---

1   GDR = 'German Democratic Republic', the name of the former socialist state in the eastern part of Germany.

**Table 10.1    German national parks**

| Year of designation | National Park | Federal land | Size (hectares) | IUCN category |
|---|---|---|---|---|
| — First phase (before 1990) — | | | | |
| 1969 | Bayerischer Wald | Bavaria | 24,250 | II |
| 1978 | Berchtesgaden | Bavaria | 20,808 | II |
| 1985 | Schleswig-Holsteinisches Wattenmeer | Schleswig-Holstein | 441,000 | II |
| 1986 | Niedersächsisches Wattenmeer | Lower Saxony | 277,00 | II |
| — Second phase (1990) — | | | | |
| 1990 | Hamburgisches Wattenmeer | Hamburg | 13,750 | II |
| 1990 | Hochharz | Lower Saxony | 6,000 | II |
| 1990 | Jasmund | Mecklenburg-Western Pomerania | 3,003 | II |
| 1990 | Müritz | Mecklenburg-Western Pomerania | 32,200 | II |
| 1990 | Sächsische Schweiz | Saxony | 9,300 | II |
| 1990 | Vorpommersche Boddenlandschaft | Mecklenburg-Western Pomerania | 80,500 | II |
| — Third phase (after 1990) — | | | | |
| 1994 | Harz | Saxony-Anhalt | 15,800 | II |
| 1995 | Unteres Odertal | Brandenburg | 10,500 | II |
| 1997 | Hainich | Thuringia | 7,610 | II |
| 2004 | Eifel | North Rhine-Westphalia | 10,700 | II |
| 2004 | Kellerwald-Edersee | Hesse | 5,724 | Not yet classified |

*Source: EPD 2006; UNEP et al. 2006*

conservation policies of the sixteen *Länder*. However, according to article 75, paragraph 1 of the Basic Law for the Federal Republic of Germany (similar to other countries' constitution), the federal government has power to enact provisions on nature conservation and landscape management as a framework for the *Länder* legislation, provided 'the establishment of equal living conditions throughout the federal territory or the maintenance of legal or economic unity renders federal regulation necessary in the national interest' (article 72, paragraph 2 of the Basic Law of the Federal Republic of Germany). The federal government has made use of its power and passed the first Federal Nature Conservation Act in 1976 which was last revised in 2002. In addition, every *Land* has its Nature Conservation Act which fills the framework established by the federal law. Protected areas are one of the major instruments of nature conservation in Germany. There are a number of other types of protected areas, including nature conservation areas, national parks, biosphere reserves, landscape protection areas, nature parks, natural monuments, protected components of landscapes and different forms of legally protected biotopes (articles

22–31 of the Federal Nature Conservation Act). Designation of protected areas falls into the exclusive competence of the *Länder*, except for the designation of national parks where the *Länder* have to consult with the federal government (article 22, paragraph 4 of the Federal Nature Conservation Act).

By law, German national parks have to be 'an entity of major size'. Additionally, the greater part of the area concerned has to be 'in a status characterized by no or little human impact' or must be 'suitable for developing/being developed into a state which safeguards undisturbed ecosystemary interactions and their natural dynamic processes to the extent possible' (article 24, paragraph 1 of the Federal Nature Conservation Act). The legal functions of national parks in Germany can be circumscribed by the terms 'natural processes', 'scientific monitoring and surveillance', 'education', and 'experience of nature' (article 24, paragraph 2 of the Federal Nature Conservation Act). Thus, German legislation is largely in line with international standards for national parks which have been defined by the World Conservation Union (see IUCN and WCMC 1994).

Most German national parks do not meet all the IUCN criteria for national parks, though. In the IUCN terminology national parks are category II. In protected areas of this type 'at least three-quarters and preferably more of the area must be managed for the primary purpose' (IUCN and WCMC 1994). One of the main purposes of national parks is, according to the IUCN, to 'exclude exploitation or occupation inimical to the purposes of designation of the area'. In practice this means that all forms of direct land-use like logging, hunting or farming would have to be eliminated on three quarters of a national park's area. This is not the case in many German national parks, including the Müritz National Park. Nevertheless all German national parks have been listed as category II protected areas in the World Database on Protected Areas (see UNEP et al. 2006).

*Marketing of National Park Regions*    Marketing of regions – or simply regional marketing – is a form of governance with the aim of stimulating processes of exchange between a region and its market partners (see Manschwetus 1995, 35). The tourism sector as one potential market of regions can exemplify that the actor constellation and the product bundle of a region usually are much more complex than in the case of an individual firm. Therefore regional marketing is characterized by cooperation of actors from different administrative levels, different jurisdictions and different sectors of society – for example politics, business, and non-governmental organizations. Another core feature is the preparation of long-term strategies in combination with easy-to-implement short-term measures. The intention is to follow a development path that integrates the desires and needs of internal and external target groups.

In management theory a distinction is made between management as an institution and management as a function (see Staehle 1992, 62–67). Similarly, there are two levels of regional marketing: a process level and a functional level. The process level includes such aspects as actors, interactions, organization, rules, and institutions. The functional level covers the contents and the achievements of regional marketing.

Ideally three consecutive stages can be observed at the process level: an initial phase, an exploration phase, and a consolidation phase. The main task in the initial phase is to reach a political decision to launch an effort of regional marketing and to find or to create an appropriate organizational nucleus. The precondition of such a decision is a common perception of a need for collective action. Public agencies can compensate the initial transaction costs by providing financial incentives.

During the exploration phase an organizational structure with at least three components ought to be established:

- A steering committee which brings together political representatives as well as high-ranking decision-makers from other sectors who dispose of relevant resources – most of all finances and power – and who can improve the overall democratic legitimation
- Several working groups in which technical experts as well as interested lay persons and citizens gather to conceive thematic strategies and projects, and
- A secretary or central coordination unit which is in charge of facilitating and managing the cooperation process.

If necessary this basic structure can be complemented by additional organizational components, for instance a coordination committee or an advisory council. Commercial communication advisors or facilitators are also frequently employed.

The exploration phase is followed by a consolidation phase in which the organizational structure often gets transferred into a more professional and stable constellation.

Generally speaking relatively little is known about the role of conservation authorities in such a marketing scheme. Therefore this is one of the central issues of the case study.

At the functional level of regional marketing, a distinction has to be made between four major steps which are interdependent in many ways:

- Analysis
- Strategy development
- Marketing mix, and
- Controlling and evaluation.

The analysis comprises an assessment of the region with its strengths, weaknesses, threats and opportunities (SWOT) as well as of the relevant markets. Once the relevant markets have been identified, the existing situation in terms of demand and supply and the major competitors can be scrutinized.

Which markets are potentially relevant for national park regions? There are four target groups in the marketing of regions in general. These are tourists, investors, residents and customers of regional export products. And what role can a national park play for these target groups? The vicinity to a national park – or more broadly speaking: scenic and unspoiled landscapes – is at best only one locational factor among many others for businesses and potential residents. However, for the tourism industry the existence of a national park can be an important factor and also for selling commercial products from the region. A national park can become an image factor for all products that benefit from being mentally associated to pureness and

**Table 10.2    Growth strategies for national park regions**

| | | MARKETS / TARGET GROUPS | |
| | | Same than before | Different or new |
|---|---|---|---|
| PRODUCTS | Same than before | **Intensification** | **Horizontal Integration** |
| | Different or new | **Innovation** | **Diversification** |

*Source: Freyer 1997, 376; Leibenath 2001, 91*

virginity, for instances food or cosmetics (see Kotler et al. 1994, 42–53; Leibenath 2001, 39, 79–84).

Strategy development deals with elaborating goals and visions which are broken down into specific action plans related to target groups and products. Strategies can not only be formulated and applied to business marketing but also to the marketing of regions. Growth strategies are one common type of strategies. If a national park region wants for example to increase its revenues from tourism, it can achieve this on the basis of different strategies. It could follow strategies of intensification, innovation, horizontal integration, or diversification (see Table 10.2). Intensification would mean that the same products are offered to the same target groups, but maybe by using the existence of the national park more extensively as a selling argument. An innovation strategy would imply to conceive new offers and products, but for the same target groups than before. The designation of a national park in an area which already is an established tourist destination could lead to such innovation. The third type of growth strategies – horizontal integration – means, that new target groups are addressed, but with the same offers than before. In this case, a national park could be used to attain the interest of other groups of customers. Finally, a strategy of diversification combines innovation and horizontal integration, which means, that new products are developed for new target groups. This would apply to regions that do only begin to develop tourism after a national park has been established there.

The marketing mix includes measures for implementing the strategies. Usually the measures are grouped into the four blocks 'product policy', 'pricing policy', 'distribution policy' and 'communication policy'.

Controlling and evaluation is needed to appraise the effectiveness and the efficiency of marketing management. It shall facilitate collective learning processes by providing information on the cost-benefit ratio. The problem here is to establish cause-and-effect-relations because the mere fact that for instance the number of tourists who visited a national park region in a certain year does not necessarily tell anything about the effectiveness of the marketing, because such an increase can have various reasons. If causal relationships are to be established, more sophisticated methods beyond sheer accounting are needed. These are mostly survey-based techniques which rely on personal interviews with a representative sample of, let's say, all visitors to a national park region (for a more detailed introduction to economic evaluation methods in nature conservation, see Leibenath and Badura 2005, 14–34).

In answering the initial question whether national park regions are a 'product' like any other, it has to be said, that the fundamental principles of marketing can

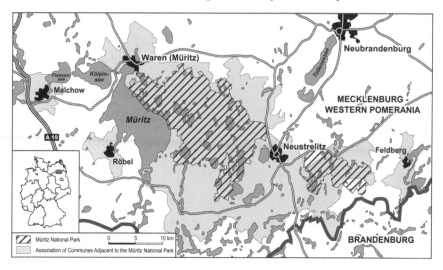

**Figure 10.1   Location of the Müritz National Park**
*Source: Drafted by author*

be applied to national park regions more or less in the same way than to any other product. On the other hand a number of particularities of national park regions can be identified, for example with regard to requirements of balancing economic concerns with a relative strict conservation regime or regarding the existence of a national park authority as a specific type of actor.

*The Müritz National Park: A Side-effect of the GDR's Collapse*

The Müritz National Park is located in the southeast part of Mecklenburg-Western Pomerania, close to the border to Brandenburg (see Figure 10.1). It has a size of 32,000 hectares and consists of two parts: One larger, western part at the shores of Lake Müritz which is Germany's second largest lake behind Lake Constance, and a smaller eastern part with an old-growth forest (see Figure 10.1) around the village of Serrahn. The national park features a richness of threatened species and biotopes, for example active raised bogs and populations of Osprey (Pandion haliaetus), Black Stork (Ciconia nigra) and Marsh Gentian (Gentiana pneumonanthe), to mention only a few (Leibenath 2001, 133). The Müritz National Park is entirely protected as a Special Protected Area according to the European Birds Directive and partly as Special Area of Conservation according to the Habitats Directive (see NPAM 2004, 21).

The national park is administered by an individual national park authority with about 150 employees. There is also an advisory board (*Kuratorium*), consisting of high-ranking local politicians, as well as representatives of public agencies, tourism organizations and other interest groups. The advisory board has to be involved in all major decisions concerning the management of the national park and must decide unanimously (see Leibenath 2001, 137).

Most parts of today's national park were used as military training area or as private hunting area of some of the GDR's socialist rulers during the post-war era until 1989. Officially, there were also about 15 smaller nature conservation areas. The Müritz region was part of the GDR National Park Programme. The regulation on the establishment of the Müritz National Park was passed in September 1990 and published in the official journal of the GDR on 1 October 1990 (see Ministerrat 1990).

Due to the unique circumstances of the political turnaround, the designation process could be handled in a very straightforward way. There were some local initiatives in favour of a national park and the first plans were discussed with local representatives in round-table talks. The whole process lasted not longer than nine months and was mainly promoted by some committed ecologists. The local population did not have a real chance to reflect the pros and cons of a national park thoroughly in those turbulent times (see Gaffert 1998, 19; Leibenath 2001, 135). This resulted in various forms of local resistance once the people had realized what it meant to live in or adjacent to a national park. Of course the national park sometimes was also pushed into the role of a scapegoat and has been made accountable for unemployment and other problems which were not necessarily caused by the national park.

Some of the communes and municipalities in and around the Müritz National Park have founded the Association of Communes Adjacent to the Müritz National Park (ACAM) – in German: *Zweckverband der Müritz-Nationalpark-Anliegergemeinden* – in 1991. One particularity of ACAM is that it joins communes from two districts (*Landkreise*): the *Landkreis* Mecklenburg-Strelitz with the capital Neustrelitz and the Müritzkreis with the capital Waren (Müritz). ACAM has the overall *raison d'être* to improve the living conditions in the region. At first it was not clear, though, whether the communes were for or against the national park. It took several years until the national park was recognized as an asset and an opportunity for the region (see Leibenath 2001, 142pp.).

The territory of ACAM is the study area for this case study and covers 840 square kilometres in total. It is characterized by poor soils. The population density of only 74 inhabitants per square kilometre, which is far below the national average, is constantly decreasing. Reversely the unemployment rate is way above the national average.

The region has a tradition in tourism that dates back to the middle of the nineteenth century. The main attractions are the lakes, streams, and extensive forests and the quietness they offer. The Müritz and the other lakes to which it is connected offer good conditions for sailing and other forms of water-based recreation. There are several small streams and rivers in the region which are excellent for canoeing and paddling. Other typical activities are hiking and biking. Especially in fall the Müritz lake is visited by thousands of cranes which attract bird-watchers.

With regard to spatial planning, the most relevant document for the case study is the regional plan (*Regionales Raumordnungsprogramm*) for the planning region Mecklenburg Lake Region (*Mecklenburgische Seenplatte*) from 1998 (RPVMS 1998). In this plan, the Müritz National Park and all other nature reserves are designated as 'priority areas for nature conservation' which means that conservation has priority against any other type of land-use. However, urban or built-up areas are excluded from this regulation and shall not be restricted in their further development. Large parts of the ACAM territory are furthermore designated as focus areas for tourism

(*Tourismusschwerpunktraum* and *Tourismusentwicklungsraum*). In these areas tourism shall have priority over other industries and the tourism infrastructure shall be improved in a way which is not harmful for sensitive ecosystems. No large-scale resorts or tourist facilities are allowed in priority areas for nature conservation like the Müritz National Park and its surroundings (see RPVMS 1998, 51pp., 98, 110).

According to the regulation on the establishment of the Müritz National Park of 12 September 1990, the national park is divided into three zones called 'I – core zone', 'II – maintenance zone' and 'III – development zone'. The purpose of zone I is to protect natural processes without disturbances whereas zones II and III serve the preservation of anthropogenic cultural landscapes. Currently zone I makes up 29 per cent of the total park area. This corresponds with a share of 3 per cent for zone II and of 68 per cent for zone III. In this context it is interesting to know that only 10 per cent of the park area is in private ownership while 64 per cent are owned by the Federal Republic of Germany, 10 per cent by the *Land* Mecklenburg-Western Pomerania, 10 per cent by communes, 1 per cent by religious communities and 5 per cent by foundations. This means that 84 per cent of the Müritz National Park is public property (see NPAM 2004: 15).

*Efforts towards Marketing of the Müritz National Park Region*

*Process Level: Actors, Framework Conditions, and Cooperation* The main actors with regard to the marketing of the Müritz National Park region are the park administration, the ACAM association, the Regional Planning Agency of the larger region to which the Müritz National Park belongs (*Amt für Raumordnung und Landesplanung Mecklenburgische Seenplatte*), a consulting firm whose representative functioned as process manager in 1996–1998, and private firms. Tourism companies like the operators of hotels and restaurants are organized in a regional branch association called Tourism Association Mecklenburg Lake Region (*Tourismusverband Mecklenburgische Seenplatte e. V.*).

As regional marketing causes both initial and permanent transaction costs and because all relevant actors were very short of funds, the dependence on external financial assistance was high. Several national funding schemes have been used between 1992 and 2005 (see BBR 2006; BMELV 2006; Leibenath 2001, 143–5):

- The federal programme 'Innovative Housing and Urban Design' (in German: *Experimenteller Wohnungs- und Städtebau/ExWoSt*), 1992–1994
- The joint programme of the Federal Government and the governments of the German Länder for the improvement of regional economies (*Bund-Länder-Gemeinschaftsaufgabe zur Verbesserung der regionalen Wirtschaftsstruktur*), 1996–1998
- The federal programme 'Regions of the Future' (*Regionen der Zukunft*), 1998–2000, and
- The federal programme 'Active Regions '(*Regionen Aktiv*), 2002–2005.

All these funds have been allocated to regional development and marketing activities.

In addition, successful project grants have been submitted under the EC programmes LEADER II and LIFE.

The sequence of funding schemes has lead to different phases of regional marketing with different actor constellations in the region. The case study is centred on the second phase from 1996 to 1998 which was dubbed Implementation Concept (*Umsetzungskonzept*), because there was common sense in the region that no more analyses and investigations were needed and that something had to happen in terms of regional development. A formal cooperation structure was established in the frame of the Implementation Concept. This structure included a steering committee which consisted of ACAM's board as well as of representatives of the national park authority and the counties (*Landkreise*). Only the members of ACAM's board had the right to vote. The steering committee established thematic working groups for the issues 'transport', 'public relation', 'visitor centres', 'food marketing', and 'sewage treatment'. The working group members were not only recruited in the public sector but also in relevant companies. The national park authority participated in those working groups which were of relevance for its mission. All meetings were held publicly.

ACAM had a secretariat which however was not equipped with full-time staff, though. That's why the consulting firm executed most of the organizational work between 1996 and 1998. This became problematic in the moment when the external financial support ended and the consulting firm could not be paid any more.

The two federal funding programmes which were the basis of the regional marketing activities in the period 1998–2000 and 2002–2005 were acquired and administered by the Regional Planning Authority.

*Functional Level: Issues and Projects* The Implementation Concept resulted in a number of projects which belong mainly to the fields of product policy and communication policy. It was problematic that the projects did not rely on a transparent analysis of the region and its relevant markets. There were some analyses which date back to the early 1990s and which remained largely unnoticed. Similarly the Implementation Concept was not based on a vision or a set of clearly defined goals. However, there was a high degree of consensus among the main actors with regard to goals. It was for example out of question that the economic capacity of the region had to be strengthened and that organic farming and ecological tourism had to be fostered. The ideas about appropriate strategies to implement these overall goals remained nevertheless fuzzy. The fact that the measures taken were not evaluated systematically was another flaw.

The so-called Müritz National Park Ticket and the creation of visitor centres around the national park are examples of measures of the category 'product policy'. The National Park Ticket was conceived in the working group on transport. Basically it was a ticket for the use of special national park bus lines, passenger ships on the Müritz lake, canoes on the rivers and of German Rail trains. Bikes were carried free of charge. The concept was flanked by a traffic management system which guides people who want to visit the national park directly from the motorway and the major highways to the bus terminals, by a system of park-and-ride areas and by a special tour programme of the national park authority. The national park ticket and the traffic management system together can be regarded as a tool for visitor management because they limited the total number of visitors and because they channelled visitors to certain spots while leaving all the remaining areas aside.

Compared to other protected areas in Germany, the Müritz National Park is very large and borders many villages and towns. Therefore it would have been difficult to establish only one central access point with one visitor centre. Against this background the idea arose to build a ring of access points around the national park. The infrastructure they offer ranges from a mere parking lot with a signpost to visitor centres with sanitary facilities, shops, and restaurants.

One example of a communicative measure were the so-called touch-boxes. These were multimedia devices which had been installed in bank offices and in visitor centres. They combined a touch-screen, a speaker, a printer and a phone booth. Their function was to provide visitors with information about the region and to allow them to make reservations in hotels or bed and breakfasts. Furthermore the national park authority was featured with an extensive coverage of the natural attractions of the park. However, the capacities of this innovative tool had not been fully utilized, because only few devices had been installed and because the interfaces to the internet were insufficient.

In the later stages of the process after 1998 a regional vision was prepared and adopted under the leadership of the Regional Planning Authority (see RPVMS 2003). In these later phases even an elaborated regional (corporate) design has been elaborated which is supposed to be used in all official written communications of the region (see RPVMS 2006). However, the national park did not play such a key role in these efforts any more like in the early phase 1996–1998. The reason is that the jurisdiction of the Regional Planning Agency is much bigger than just ACAM's territory and that it contains two other large protected areas which have the status of 'nature parks'.

*Conclusions for the Integration of Conservation and Competitiveness in National Park Regions*

*Sharing Tasks between Conservation Authorities and Other Actors* The case study shows the importance of establishing a coordination unit or secretariat as the basis of a regional marketing initiative. Such a secretariat can also be transformed into a permanent development agency at the transition from the exploration phase to the consolidation phase. At first glance it might be tempting to put a national park authority in charge of this task as for instances the administration of the Müritz National Park has the explicit function to contribute to the economic development of the region (article 3 of the regulation on the establishment of the Müritz National Park). However, a conservation agency will inevitably reach the limits of its credibility if it promotes a very rigid form of conservation and at the same time tries to boost the regional economy.

The 'owner' of a regional marketing initiative should be an organization from outside a national park, because marketing of national park regions mostly refers to activities beyond the protected area. In the case of the Müritz National Park, ACAM met this criterion and had also the advantage that its members disposed of the buildings which were needed to create the ring of access points. However, there was also the problem that the members of the association were quite heterogeneous with regard to their size, resources, and interests. Moreover the members were public agencies which were subject to fiscal accounting rules which meant that their ability to operate as drivers of regional development is much more restricted than those of an intermediate or private organization.

Other than one might expect, the involvement of the Müritz National Park Authority in the regional marketing initiative did not provoke any resistance within the administration, because staff and superior authorities as well had an interest in making the national park a positive factor for regional development.

The involvement of the national park authority is a precondition without which a national park region hardly can be developed in a way that makes full use of the park's potentials without conflicting with the conservation goals. On the other hand the national park authority should consider the interests of other regional actors in decisions about the management of the park in order to follow a consistent development path in the entire region. Such a system of mutual participation can be reached best through the working structure of regional marketing. In the Müritz region, additional linkages were created by including different groups of actors in the park's advisory council. In reverse the national park authority is member of the regional tourism association. Thus the national park authority knows the other actors's interests and can consider them, for example when it comes to modifications of the trail network within the park. And the park authority participates in decisions outside its jurisdictions, for example in transport affairs.

*Finding the 'Right' Issues at the Beginning and Mobilizing Financial Assistance*

The issues of transport systems and tourist information have proven to be very well suited for the initial and exploration phases of regional marketing. They offer the big advantage to start with the region's strengths instead of its weaknesses. Otherwise it can easily be the case that responsibility is delegated to external organizations.

The Müritz case presented in this chapter did only cover a relatively short period of time within a longer history of approaches towards regional development in and around the Müritz National Park. By looking at this broader picture some more general observations can be made.

The Müritz National Park is located in a peripheral region which is economically weak, very sparsely populated and characterized by a lack of associational structures between actors and organizations across their respective jurisdictions. In such regions, regional market initiatives require long-term financial assistance from outside in order to build up institutional capital and effective governance relations. However, this can only be achieved as the outcome of historical processes because associational structures imply also informal conventions, routines and habits which are embedded in the locality and sustained over time. However, a national park or other types of large protected areas can help to generate a shared place-based identity amongst the partners – especially in cases like the Müritz National Park region with historically and politically separate entities and without a single administrative authority responsible for the whole area (see Davoudi 2003, 992; Fürst et al. 2005, 332).

*Defining the Scope of the Region to be Marketed*

The Müritz case shows that it is not necessarily clear from the outset where the boundaries of a national park region are. In fact, there can be a number of overlapping cooperation spaces, depending on the issue, the orientation of relevant actors, and outside stimuli. In the Müritz case, the regional marketing initiative started in the mid-1990s only on the territory of the ACAM association. Later on it merged into the larger region, that is the jurisdiction of the regional planning authority who did participate in several national programmes for regional development.

The bottom line is, however, that the shape of the cooperation space around the Müritz National Park has been defined neither on the basis of functional relations and interdependencies nor as a consequence of spatial identities derived from the park. Actually it was the result of the interplay between national funding structures and the successful application for these funds which have been submitted by committed individuals. To be more specific: Because the head of a small consulting firm and the director of the Müritz National Park submitted a grant proposal for the marketing of the ACAM region, the so-called Implementation Concept was focused on ACAM's territory in the years 1996–1998. And because one of the chief administrators of the regional planning authority was successful in applying for other national funds in the following years, the focus shifted to the larger jurisdiction of this organization.

There are a few more things to be learned by looking at the changing shape of the cooperation area in the Müritz case. It makes sense to draw the line not too narrowly around the protected area but to define a cooperation space that can work as a tourist destination. Most tourists don't visit European regions only because of a protected area and the natural beauty it preserves, even if there might some exceptions from the rule. Instead, they are looking for a broad mixture of outdoor and indoor recreation facilities as well as for opportunities for cultural and sportive enjoyment. As a national park should be prevented from turning into a sort of Disneyland, it is wise to define functional regions for regional marketing relatively large so that the attractions from various towns and villages can be integrated (see Leibenath 2001, 96–99).

At the same time it is advisable not to orientate the delineation at functional criteria alone. For an effective evaluation of regional marketing activities, statistical data are needed, for instance on the number of tourists or on the shares of different business sectors in the regional economy. Usually such statistical data are not collected for smaller communes individually but only for larger territorial units. The better availability of statistical economic data was a positive side-effect of lifting the marketing activities of the Müritz National Park Region on the level of the Regional Planning Agency which includes three *Landkreise*.

Marketing is about competitiveness. In this sense the marketing of national park regions can help to get away from the concentration on what is feasible in terms of regional development and to focus attention on what is needed to meet the desires and interests of relevant target groups. By doing so the legitimacy of a national park can be raised among local stakeholders and at the same time the attractiveness of the entire region for tourists, customers and investors can be improved. Thus, marketing is a promising approach to the governance of biodiversity which can compliment and substantiate other concepts.

## References

BBR (= Bundesamt für Bauwesen und Raumordnung (2006), *Netzwerk Regionen der Zukunft. Informationen zum Modellvorhaben]*, available online: http://www. zukunftsregionen.de/informationen/main.htm, 02.02.2006.

BMELV (= Bundesministerium für Ernährung, Landwirtschaft und Verbraucherschutz (2006), *Regionen Aktiv – Was ist Regionen Aktiv?*, available online: http://www. nova-institut.de/modellregionen/, 02.02.2006.

Davoudi, S. (2003), Polycentricity in European Spatial Planning: From an Analytical Tool to a Normative Agenda, *European Planning Studies* 11(8), 979–999.

EPD (= Europarc Deutschland) (2006), *Deutsche Nationalparke*, available online: http://www.nationalparke.de/pages/parke/natpark.htm, 30.01.2006.

European Union (2000), *Presidency Conclusions. Lisbon European Council. 23 and 24 March 2000*, available online: http://ue.eu.int/ueDocs/cms_Data/docs/pressData/en/ec/00100-r1.en0.htm, 06.06.2005.

Freyer, W. (1997): *Tourismus-Marketing: Marktorientiertes Management im Mikro- und Makrobereich der Tourismuswirtschaft.* (München, Wien, Oldenbourg).

Fürst, D., Lahner, M. and Pollermann, K. (2005), Regional Governance bei Gemeinschaftsgütern des Ressourcenschutzes: das Beispiel Biosphärenreservate, *Raumforschung und Raumordnung* 5/2005, 330–339.

Gaffert, P. (1998), Akzeptanzprobleme in Großschutzgebieten in: Erdmann, K.H., Wiersbinski, N. and Lange, H. (eds), *Zur gesellschaftlichen Akzeptanz von Naturschutzmaßnahmen – Materialienband.* (Bonn, Bundesamt für Naturschutz), 19–20.

Getimis, P. and Heinelt, H. (2004), *Leadership and community involvement in European Cities. Conditions of Success and/or Failure*, available online: http://www.uic.edu/cuppa/cityfutures/papers/webpapers/cityfuturespapers/session8_4/8_4leadershipcommunity.pdf, 16.09.2005.

IUCN (= The World Conservation Union) & WCMC (= World Conservation Monitoring Centre) (1994), *Guidelines for Protected Area Management Categories*, available online: http://www.unep-wcmc.org/index.html?http://www.unep-wcmc.org/protected_areas/categories/eng/~main, 30.01.2006.

Jungius, H. (1985): Das Nationalparkkonzept heute und im Rahmen der internationalen Entwicklung, in Arbeitsgemeinschaft beruflicher und ehrenamtlicher Naturschutz e. V. (ed.), *Nationalparke: Anforderungen – Aufgaben – Problemlösungen.* (Greven: Kilda), 9–17.

Kotler, P., Haider, D. and Rein, I. (1994), *Standort-Marketing: Wie Städte, Regionen und Länder gezielt Investitionen, Industrien und Tourismus anziehen.* (Düsseldorf, Vienna, New York, Moscow, Econ).

Leibenath, M. (2001), *Entwicklung von Nationalparkregionen durch Regionalmarketing, untersucht am Beispiel der Müritzregion.* (Frankfurt, Peter Lang).

Leibenath, M. (in print), Legitimacy of Biodiversity Policies in a Multi-Level Setting – the Case of Germany, in Korthals M. (ed.), *Multilevel Policy Making in a Democratic Context: European Nature Conservation Policy, its Local Implementation and the Growing Salience of Legitimacy.* (Berlin, Springer).

Leibenath, M. and Badura, M. (2005), *Manual for the Evaluation of Natura 2000 Sites in Economic Terms,* available online: http://www.ioer.de/PDF/PublikPDF/Natura2000_engl.pdf, 30.01.2006.

Manschwetus, U. (1995), *Regionalmarketing: Möglichkeiten und Grenzen des Marketing-Managementansatzes als Instrument der Regionalentwicklung.* (Berlin, Free University of Berlin).

Möller, C. (1995), Strategien der Naturschutzverwaltung und Naturschutzpolitik: Vom Anwalt zum Mediator, *Politische Ökologie* 43, 28–32.

NPAM (Nationalparkamt Müritz) (2004), *Nationalparkplan,* available online: http://www.nationalparkamt-mueritz.de/?id=10&file=nationalparkplan&lang=de, 12.10.2005.

RPVMS (Regionaler Planungsverband der Planungsregion Mecklenburgische Seenplatte) (1998), *Regionales Raumordnungsprogramm Mecklenburgische Seenplatte.* (Neubrandenburg, RPVMS).

RPVMS (Regionaler Planungsverband der Planungsregion Mecklenburgische Seenplatte) (2003), *Leitbild der Region Mecklenburgische Seenplatte [Vision for the Mecklenburg Lake Region],* available online: http://www.region-seenplatte.de/Leitbild/Leitbild.pdf.

RPVMS (Regionaler Planungsverband der Planungsregion Mecklenburgische Seenplatte) (2006), *Designhandbuch. Regionales Design Mecklenburgische Seenplatte,* available online: http://www.seenplatte-mueritz.de/index.php?&seiten_id=201.

Runte, A. (1987), *National Parks. The American Experience.* (Lincoln, London, University of Nebraska Press).

Scharpf, F.W. (1998), *Interdependence and Democratic Legitimation,* available online: http://www.mpi-fg-koeln.mpg.de/pu/workpap/wp98-2/wp98-2.html, 09.09.2005.

Scharpf, F.W. (2004), *Legitimationskonzepte jenseits des Nationalstaats,* available online: http://www.mpi-fg-koeln.mpg.de/pu/workpap/wp04-6/wp04-6.html, 09.09.2005.

Staehle, W.H. (1992), *Funktionen des Managements.* (Bern, Stuttgart, Haupt).

UNEP (United Nations Environment Programme), WCMC (World Conservation Monitoring Centre) and WCPA (World Commission on Protected Areas) (2006), *World Database on Protected Areas,* available online: http://sea.unep-wcmc.org/wdbpa/, 31.01.2006.

Wolf, K.D. (2002), *Civil Society and the Legitimacy of Governance Beyond the State - Conceptual Outlines and Empirical Explorations,* available online: http://www.isanet.org/noarchive/wolf.html, 16.09.2005.

## Legislation Documents

Basic Law for the Federal Republic of Germany from 23 May 1949, (Federal Law Gazette, p. 1) (BGBl III 100-1), most recently amended by the amending law dated 26 July 2002 (BGBl I, p. 2863), available online: http://www.bundesregierung.de/static/pdf/GG_engl_Stand_26_07_02.pdf, 17.10.2005.

Federal Nature Conservation Act of 25 March 2002 [Official Translation], available online: http://www.bmu.de/files/pdfs/allgemein/application/pdf/bundnatschugesetz_neu060204.pdf, 12.10.2005.

Ministerrat (= Der Ministerrat der Deutschen Demokratischen Republik [Council of Ministers of the German Democratic Republic] & Amt des Ministerpräsidenten [Office of the Prime Minister]) (1990), Verordnung über die Festsetzung des Nationalparkes 'Müritz-Nationalpark' vom 12. September 1990 [Regulation on the Designation of the National Park 'Müritz National Park' of 12 September 1990], *Gesetzblatt der Deutschen Demokratischen Republik*, Sonderdruck Nr. 1468, 8.

## Chapter 11

# A Future Model for Protected Areas and Sustainable Tourism Development: The New National Parks in Scotland

L. Rory MacLellan

### Introduction

The links between protected areas, in particular national parks, regional development and tourism are long established but problems associated with increasing demands of tourism are testing park management to the limit. Reconciling the core environmental protection aim with social, cultural and economic pressures, associated with tourism growth, becomes ever harder to achieve. Conflicts arise over contested space, issues of costs and benefits and increasingly who foots the bill. Transport congestion, regulations, controls and draconian pricing threaten to diminish the intrinsic qualities of protected areas and their enjoyment by the public. The sustainable development paradigm applied to tourism offers frameworks that may help. Scotland provides an interesting case study of this application for three reasons: first, it has adopted a broad interpretation of the national park designation, similar to the established UK model of multiple owned, 'lived in' parks; second, the parks are very recent and as such are still evolving; third, they purport from the outset to incorporate sustainable development objectives. This last point is demonstrated by the extent to which sustainable tourism lies at the core of park strategies.

The chapter examines an interesting new model that might give insight into the core issue of reconciling protection of nature and regional development. The model for national parks in Scotland has at its core a holistic approach derived from sustainable development philosophy that includes environmental conservation and recreational goals along with those of economic and social development. Although in some respects similar to the rest of the United Kingdom, the peculiarities of Scotland's complex relationship with land and sensitivities surrounding protection of the cultural heritage partly explain why it has taken so long to establish national parks and the broad perspective of the model adopted. The so-called wilderness areas of Scotland may seem peripheral and sparsely populated but have in fact been inhabited for generations. True wilderness areas, untouched by human activities are rare. Depopulation is a relatively recent phenomena and the economic imperative has always been a priority, especially in the Highlands, in order to anchor fragile communities and thereby preserve their cultural heritage.

Therefore the chapter reviews the new model for protected areas that incorporates broader sustainable development priorities. The focus is on Scotland's recently constituted national parks and the implementation of sustainable tourism strategies. The background and unique characteristics of the National Park model in Scotland are examined in addition to alternative approaches to achieving sustainability through tourism. The chapter argues that the European Charter for Sustainable Tourism (ECTS) provides a useful framework and context, in particular for The Cairngorms National Park.

The case study focuses on The Cairngorms, designated as a national park in January 2003, and becoming Scotland's second and the UK's largest national park. The fourth aim of Scotland's national parks as laid out in the National Parks (Scotland) Act 2000, 'to promote sustainable social and economic development of the communities of the area', provides particular justification for the park authority to become actively engaged in the development and promotion of sustainable tourism. However, it is recognized that well-managed tourism should also significantly contribute to each of the other three aims: 'to conserve and enhance the natural and cultural heritage; to promote the sustainable use of the natural resources of the area and to promote understanding and enjoyment of the special qualities of the area by the public'.

The Cairngorms National Park Authority has sought to reap the benefits of sustainable development whilst ensuring that the natural and cultural heritage resources are enhanced and protected. It is in the process of developing an effective framework for planning, action and evaluation for sustainable tourism that encompasses the many existing tourism initiatives and programmes but also incorporates an international dimension, hence the adoption of ECTS. As part of this process they have established a private sector led Tourism Development Working Group that includes representatives from all key organizations with responsibility for tourism in the Cairngorms. The process of working towards Charter status has raised several interesting issues.

**Tourism, Sustainability and National Parks**

Sustainable development theory is based on the concept that environmental protection and economic growth can be compatible objectives (see Hardy et al., 2002). Tourism has been gradually incorporated into this concept. The goal of sustainable tourism is for all tourism activities, regardless of scale or location, to be part of the sustainable development agenda. Many principles and guidelines have been developed for sustainable tourism in attempts to include environmental, cultural, economic and social goals within the context of tourism. For example, the acronym VICE represents four sustainable tourism aims: visitor satisfaction, industry profitability, community acceptance, and environmental protection (see Stevens 2002). Sustainable tourism has particular resonance for national parks and a protected area where nature is fragile and ecotourism or Nature Based Tourism (NBT) relies on the long term well being of the environment (see MacLellan 2001). Furthermore, the concept offers great utility in how parks may be planned and managed for tourism (see Boyd 2002).

Throughout history there have been examples of efforts made by governments and landowners to protect areas with special natural attributes for their intrinsic value and recreational qualities. Government involvement and responsibility for landscape protection has become the norm and government is now seen as the primary delivery vehicles of protected areas. Some argue that this involvement has gone too far in terms of complexity and number of protected area designations (see Bishop et al. 1997). However it is the growth of national parks over the last ten years that has been remarkable. Eagles (2002) estimated there were 30,361 parks and protected areas in 1996 and in 2002 the number of national parks alone had risen to 3,386 worldwide. National parks are viewed as the top tier designation and their scale and attractions have always acted as a magnet for tourism activities, resulting today in the search for sustainable solutions to reconcile tourism and conservation demands.

An international classification of protected areas, established by the International Union for Conservation of Nature and Natural Resources (IUCN), includes scientific nature reserves; national parks; natural landmarks; nature conservation reserve; protected landscapes; resource reserve; natural biotic area; multiple-use management area; biosphere reserve and world heritage sites. The majority of National Parks follow the American model and come under category II of the IUCN classification where land is largely uninhabited, publicly owned and access is controlled (see Leitmann 1998, 129). Despite this conservation priority and strict regulations, from an early stage national parks had a clear connection between the park, tourism and recreation. Issues arising from attempts to reconcile growing pressures from late twentieth century tourism demand with the conservation priority are testing the category II model to the limits (see Boyd and Butler 2000). Britain, on the other hand, is unusual in that areas designated as national park come under category V of the IUCN classification mainly because areas are populated, privately owned and arguably require a holistic, or sustainable, approach based on social, environmental and economic issues. This in large part is a function of island restrictions and early industrialization leading to high population densities and the need for maximum resource exploitation. Even seemingly wild, natural land in the north of Scotland has experienced intensive use at one point. Thus Category V protected landscape/ seascape has been more appropriate, defined as 'an area of land, with coast and sea as appropriate, where the interaction of people and nature over time has produced an area of distinct character with significant aesthetic, ecological or cultural value, and often with high biological diversity' (IUCN 1994).

Although these parks share many tourism development pressures experienced by their more 'pure' category II national park cousins it could be argued that they are better placed to deal with twenty first century pressures. Today everybody wants a piece of the park, in particular local communities that recognize the potential development value the designation brings. The regional development imperative has been acute in Western European peripheral areas resulting from declining values in traditional agricultural produce. Tourism has been promoted as one means of economic diversification and protected areas often form a key part of the destination visitor attraction mix. Other examples include parks experiencing escalating numbers of stakeholder conflicts relating to distribution of costs and benefits, destination image and promotion, managing nature in parks, access and sustainable development. This is compounded by increasing visitor numbers and diversity of visitor expectations and behaviours. All exert pressures on protected area management. The growth in emphasis on wider stakeholder involvement in park decision making continues to complicate policy-making, with demands to include broader, sustainable development objectives as opposed to traditional environmental conservation ones. In a study of Banff National Park, Canada, the authors note: 'The management of protected areas must increasingly contend with the philosophical debate of use versus preservation, as urbanization, modernization, population mobility and international tourism growth continue to impact diminishing and fragmented green spaces' (Jamal and Eyre 2003, 417).

The growth of specialist tourism segments, such as nature based tourism (NBT), often takes place in national parks and despite its benign image, the volume and specialized activities still manage to exert environmental pressures (see Laarman and Gregersen 1996, Boyd and Butler 2000, Eagles 2002). This movement raises many questions relating to funding, access, transport, carrying capacity, visitor management and pricing (see WTO 1992). The issues of willingness to pay for access to parks and community costs and benefits become acute with rapid growth and associated impacts. While measuring costs and benefits is more commonplace in the clearly delineated and closely controlled North American parks their structures are less adaptable to change. The category II model has multiple objectives built in and recent studies have adapted similar methodologies. Liston-Heyes and Heyes (1999) use the travel cost method in Dartmoor, England to evaluate benefits users derive from access to the park.

Yet solving funding issues in current parks remains elusive as visitor numbers grow disproportionately to the levels of funding, visitor fees and user charges, creating a situation where increased maintenance and refurbishment requirements influence the quality of services (see van Sickle and Eagles 1998; Eagles 2002; Buckley 2003). National park authorities need to review their pricing policy, which affects entrance fees and other charges, and reflects the true 'willingness-to-pay' (Laarman and Gregersen 1996). Much of the literature also indicates the importance of implementing a pricing framework consisting of a mixture of regulation and incentives, 'sticks' and 'carrots' (see Leitmann 1998, van Sickle and Eagles 1998, Eagles 2002, Buckley 2003). The situation in Scotland presents an opportunity to examine a, perhaps, more holistic approach to resolving these issues. Rather than simply searching for means to pay for the costs of protecting natural areas from the national purse, local taxation or visitor charging schemes, the challenge has been turned around to identify ways in which regional economies in and around protected areas can gain direct benefits. Careful, innovative local management of tourism in Scotland's national parks presents a critical prospect.

**Scotland's National Parks**

The sustainable development paradigm has been adopted by government in Scotland for over a decade. Tourism is a key economic sector for Scotland and should be planned sustainably: ecologically bearable; economically viable; and ethically and socially equitable for local communities. Relevant Scottish agencies share common sustainable development aims but at individual project or site level this conflict between development and conservation interests still arise. Co-ordination of policies has been assisted by the establishment of partnerships such as the Tourism and Environment Forum (TEF) involving a wide spectrum of public, private and voluntary interests. Work on raising industry environmental awareness and integrating environmental variables into tourism quality grading schemes (Green Tourism Business Scheme) has been slow and local sustainable tourism pilot projects (Tourism Management Programmes) were useful test-beds but most did not last. Sustainable tourism, it seems, had slipped off the agenda in national tourism policy documents indicated; 'A New Strategy for Scottish Tourism' (Scottish Executive 2000); 'Tourism Framework for Action: 2002 – 2005' (Scottish Executive 2002). However the latest tourism strategy 'Scottish Tourism: the next decade – a tourism framework for change' (Scottish Executive 2006) has to some extent redressed the balance by placing sustainable development as a core objective of tourism strategy.

Protected areas and NBT are key testing grounds of the implementation of this policy: recent progress with legislation on access to the countryside is useful however issues of 'who pays for countryside maintenance' remain unresolved. Charging entrance to parks is not possible in the UK so a combination of balancing land management regulations, visitor management and promotions and taxation/ subsidization are applied. As tourism depends on a mixture of private and public provision of attractions and facilities, private landowners, conservation organizations and local authorities question whether those who come and enjoy these features are contributing their proper share of the costs involved (see Parker and Ravenscroft 1999). The new national parks offer opportunities to combine economic benefits to the locality from sustainable tourism while minimizing environmental impacts.

As discussed, Britain's late adoption of national parks in 1949 is attributed to the predominance of humanized landscapes so the model chosen lacks coherence with the IUCN category V. The national parks in England and Wales are characterized by a high degree of privately owned land, living communities and neighbourhoods and fewer access restrictions. The opportunity to exploit land for consumptive practices such as tourism is hence much greater. The majority of conflicts in British national parks can thus be linked to this (see Parker and Ravenscroft 1999). Success in attracting tourists, the majority in private cars, has led to problems of congestion, pollution, erosion, litter, and land use conflict leading to calls for more sustainable tourism development (see Lovelock 2002; Laarman and Gregersen 1996).

Scotland shares similar land ownership characteristics with the rest of the UK however national parks were rejected in 1949 due to a combination of opposition from landowners fearing land nationalization and local authority concerns with further depopulation evoking memories of the 'highland clearances'. In addition, economic development through the in vogue hydro electric schemes was given top priority (see Warren 2002). Finally, the need for public recreation was less urgent than in England (see Moir 1997).

So, despite Scotland having natural heritage characteristics more suited to IUCN category V national parks, for fifty years it remained one of the few countries without any form of national park. The following forty years saw an unprecedented growth in demand for the Scottish countryside. By the late 1980s, the weak land management arrangements were straining under a series of highly publicized tourism related conflicts, resulting in lengthy consultation, and finally recommendations for national parks in the Cairngorms, Loch Lomond and the Trossachs, Ben Nevis-Glen Coe-Black Mount, and Wester Ross. 'The proposals included independent planning boards comprising local and national members' (McCarthy et al. 2002, 668).

Public support was overwhelming however the government of the day thought national parks unsuitable, suggesting alternative voluntary partnership arrangements. Partnership boards for the two high profile areas were set up but arguably without the necessary powers or funding to cope with growing development pressures. The true sustainable development credentials of these policies may be judged against international criticism such as a World Conservation Union report which condemned Scotland for: 'operating one of the weakest management arrangements for vulnerable areas in Europe' (Edwards et al. 1993, 6).

Campaigns for national parks continued however the change came abruptly with a change of government in 1997. Two announcements came in quick succession from Donald Dewar, Secretary of State for Scotland: first devolution for Scotland, 'there shall be a Scottish Parliament'; secondly the announcement of the Government's commitment to the establishment of national parks.

> I believe that National Parks are the right way forward for Scotland. The major gap we have identified in the current system of natural heritage designations relates to the management of a small number of relatively large areas of natural heritage importance, of which Loch Lomond and the Trossachs is a prime example. What we require in these areas is an integrated, rather than sectoral, approach to their management (Scottish Office 1997).

The detail in the announcement was significant. After years of consultation and debate over alternative designated area models for Scotland, in one political gesture, the decision was made: 'Instead of proceeding logically from problem to diagnosis to prescription the political decision that Scotland must have national parks came first, and only then was attention given to what their form and function should be' (Warren 2002, 213).

The statutory agency responsible for implementation had to catch up fast. In 1998, Scottish Natural Heritage (SNH) launched the main consultation paper 'National Parks for Scotland' and the Government formally accepted this report in 1999. The Loch Lomond and the Trossachs National Park Interim Committee was established in 1999 to pave the way for the national park and in particular develop the strategic thinking necessary for the early preparation of a national park plan and establish effective partnership working across all areas of interest. Following the passing of the National Parks (Scotland) Act in 2000, Scottish Ministers made a formal proposal for a national park in the Loch Lomond and Trossachs area. Loch Lomond and the Trossachs National Park formally opened on 24 July 2002. The Cairngorms National Park followed on 1st September 2003 (see Figure 11.1).

The details of the enabling legislation are worth closer examination. The National Parks (Scotland) Act 2000 makes provision for national parks. Section 1 of the Act lists the core aims of Scotland's national parks as:

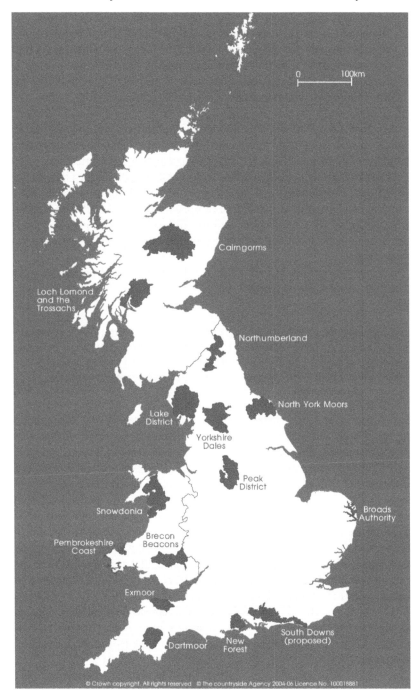

**Figure 11.1   National parks in England, Wales and Scotland**
*Source: The Countryside Agency 2004-06 Licence No. 100018881. http://www.*
*nationalparks.gov.uk/.*

1. To conserve and enhance the natural and cultural heritage of the area
2. To promote sustainable use of the natural resources of the area
3. To promote understanding and enjoyment (including enjoyment in the form of recreation) of the special qualities of the area by the public and
4. To promote sustainable economic and social development of the area's communities (see HMSO 2000).

It is clear that these aims go beyond traditional objectives for national parks and are based on core sustainable development themes of social and economic aims in addition to environmental. Aims (c) and (d) illustrate two main differences with this model and explain why Scottish national parks comes under category II of the IUCN protected area classification and not category V. The ability to promote national parks is highly unusual and is not part of comparative English and Welsh legislation. The responsibility for the economic and social development of park communities is also unusual and linked to the British tradition of designating national parks in populated areas.

The Government was heavily criticized for aim (d) and decided to include the Sandford Principle to strengthen the conservation aspect. Section 9, point (6), states:

> In exercising its functions a National Park Authority must act with a view to accomplishing the purpose set out in the subsection (1); but if, in relation to any matter, it appears to the authority that there is a conflict between the National Park aim set out in section 1(a) and other National Park aims, the authority must give greater weight to the aim set out in section 1(a) (HMSO 2000).

A pivotal issue in the debates in the Scottish Parliament related to the balance between conservation aims and social and economic development. Responses to consultation highlighted concerns that social and economic issues were to take second place to conservation. Drawing park boundaries to include less developed rural communities exacerbated this. Consequently, the bill was amended to ensure that the National Park Authorities (NPAs) accept an integrated approach in order to reconcile competing interests. It was established that a national park authority would only be required to give greater weight to conservation after failing to resolve a conflict. The emphasis placed on sustainable development indicates they 'are not intended to be preserved areas in which all development is fossilized; instead, they are intended to be places that set an example of how to integrate the rural economy with the protection of the natural and cultural heritage' (McCarthy et al. 2002, 669).

It is still too early to judge whether this represents a modern, innovative framework for national parks, a model for sustainability in the twenty-first century or a watered down designation based on an already weak British interpretation of a national park.

Tourism has been identified as a critical factor in securing environmentally and socially sustainable development for areas within the two national parks. The onus is therefore on park authorities, together with their partners and stakeholders, to avail themselves of good practice in sustainable tourism. Whilst it would appear that sustainable tourism should contribute most to the delivery of the all-important fourth aim 'to promote sustainable economic and social development of the area's communities', the development and management of sustainable tourism has the potential to contribute to all four primary aims. As noted earlier, Schedule 3 of the Act sets out specific powers for the NPAs with respect to tourism. This includes the provision of information, education and interpretive facilities as well as services to promote the enjoyment of the parks' environments. Importantly and significantly, the NPAs are able to provide tourism facilities in the national parks and encourage persons to visit the national parks (paragraph 4, Schedule 3 of the Act). This ability to directly provide, manage and intervene creates interesting opportunities to be proactive, especially in partnership with other stakeholders, to provide sustainable tourism programmes. Unlike NPAs in England and Wales, the Scottish NPAs will be able to 'provide, or encourage other persons to provide, facilities in the National Park; and encourage persons, by advertisement or otherwise, to visit the National Park' (National Park (Scotland) Act 2000, paragraph 4, Schedule 3 of the Act). This additional power gives the Scottish NPAs the chance to engage in creative destination marketing to promote sustainable tourism.

In addition, SNH sets out a vision for parks that includes the following key elements relevant to sustainable tourism development: national parks should engender trust between national and local interests in the delivery of conservation and community objectives; and national parks should be pioneers of techniques for achieving sustainable development (see SNH 1999, 8).

### Combining Regional Development Strategies with Tourism and National Park Plans

Partnership working and research into sustainable tourism strategies for Scotland's national parks began before the parks themselves were established. Scotland was keen to devise unique models and draw on international best practice. For example, prior to the setting up of the Cairngorms National Park Authority (CNPA) a tourism forum, the Tourism Development Working Group (TDWG) was convened, comprising representatives from all tourism interests within the Cairngorms National Park area. An extensive consultative process, building on decades of debate within the region has, it seems, finally brought about the consensus to build a management mechanism balancing conservation and development that overcomes historical scepticism and opposition to the idea of a top-down protected area designation for Scotland's upland areas.

Two important and highly relevant pieces of research were drawn on: 'Sustainable Tourism in National Parks and Protected Areas' (Stevens 2002) and 'The European Charter for Sustainable Tourism in Protected Areas – A Prospectus for Action in Scotland's National Parks' (EUROPARC Consulting, 2003). A series of preparatory reports covering a range of subject areas were commissioned by Scottish Natural Heritage (SNH) in advance of the national park, and sustainable tourism was seen as an important area where significant information could be collated on best practice examples throughout Western Europe. In it's 1999 'Advice to Government' on national parks for Scotland SNH set out a vision for parks that included the following key elements relevant to sustainable tourism development (Scottish Natural Heritage 1999, 8): 'national parks should engender trust between national and local interests in the delivery of conservation and community objectives; and national parks should be pioneers of techniques for achieving sustainable development'. The Stevens Report 2002, had three objectives: to identify the key principles for sustainable tourism in Scotland's national parks, to illustrate these principles with a number of case studies, and to make recommendations on arrangements for collaborative working. It reviewed, analysed and evaluated best practice in respect of: the current guidelines for tourism in protected areas in Britain and elsewhere; and management best practice in terms of policy and planning; monitoring and review of impacts; use of facilities provided by managing bodies and the range of strategies, tools and techniques for promoting sustainable tourism. As a result the report provides both specific examples of sustainable tourism in action and highlights the importance of appropriate policies within the NPAs to enable and facilitate sustainable tourism development.

The research into the European Charter for Sustainable Tourism in protected Areas in Scotland was jointly commissioned by Loch Lomond and The Trossachs National Park, the Cairngorms Partnership (the organization responsible for the management of the Cairngorms area prior to the set of the national park) and SNH. All partners were familiar with the Charter through presentations and information from Europarc and felt that further research into the potential suitability for Scotland's future national parks was worthwhile. A joint approach would also build on the existing collaborative visitor survey work. The Europarc Consulting Report (2003) identified the structures and activities relating to sustainable tourism at that time in each park area and looked at ways in which these might need to be strengthened in order to meet the requirements of the Charter. It identified seven benefits for the parks and these included raising the profile of the parks and sustainable tourism and providing an opportunity to align the policies to current international thinking and practice; networking with other Charter parks and the helpful internal and external assessment. In summary, the report noted that it would be important not to lose momentum for this area of work in the setting up of the national parks.

## Case Study on Cairngorms National Park

The Cairngorms area is a national asset for Scotland and the UK. It is an area that has long been considered a special place for its natural and cultural heritage, as a place to live, work and visit. Tourism has existed here since the nineteenth century with particular associations with field sports and royalty (Balmoral and Royal Deeside) and in the latter half of the twentieth century, winter sports. The Cairngorms area lies in the central highlands of Scotland and is centred on the large massif of the Cairngorm Mountains, a unique area of high ground in the UK. The area is easily accessible from many areas of Scotland and has good communications from the major nearby centres of population, including Inverness, Aberdeen, Dundee and Perth and is under two hours from the major conurbations of Edinburgh and Glasgow. The area also satisfies conditions set out in legislation in order to be designated as a national park in Scotland:

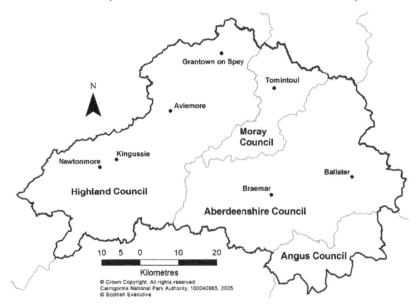

**Figure 11.2   Cairngorms National Park local authority areas**
*Source: Cairngorms National Park Authority 2005*

- That the area is of outstanding national importance because of its natural heritage, or the combination of its natural and cultural heritage
- That the area has a distinctive character and a coherent identity
- That designating the area as a national park will meet the special needs of the area.

The designated park area has a population of some 16,000 people living and working in communities in parts of the Highland, Moray, Aberdeenshire and Angus Council areas. Although the designation as a national park will bring more cohesive land use management to the area, full planning powers remain with the above local authorities. Each authority takes a slice of the park but their constituencies are essentially focused on major town and population centres outside (see Figure 11.2). The park is centred on the Cairngorm Mountains and extends to Grantown-on-Spey, Strathdon, Ballater, the heads of the Angus Glens, Dalwhinnie and Laggan. The park has a much lower population density than Scotland as a whole. Occupying an area of 3,800 square kilometres, or 408,782 hectares, it has a population density of just 0.04 people per hectare. This compares with a Scottish average of 0.65 people per hectare. The park is host to around 500,000 staying visitors each year and a far greater number who come for the day or pass through (see CNPA 2006).

   The pattern of employment is more typical of a peripheral rural area and reveals an economy which is relatively narrow compared with that for Scotland as a whole. There are a small number of sectors that are over-represented and a larger number where representation is below average. However, this pattern is common to much of the Highlands and Islands. A higher proportion of the park population (5.7 per cent)

**Table 11.1    Top 12 features most liked by visitors to the Cairngorms area**

| Cairngorms features identified by visitors | Percentage of respondents |
|---|---|
| Beautiful views and scenery/spectacular | 46% |
| The hills/wide spaces, rugged mountains | 27% |
| Peacefulness and easy-going pace of life | 25% |
| The trees and colours of the landscape | 13% |
| The wildlife, plants and animals | 11% |
| Nice walks, good hill-walking | 11% |
| Fresh, clean unspoilt area | 11% |
| Friendly people | 9% |
| Picturesque, very beautiful place | 9% |
| Lots of things/activities to do | 9% |
| Large, open spaces without seeing anyone | 7% |
| The water, lochs and waterfalls | 6% |

*Source: Cairngorms National Park Visitor Survey 2004*

are employed in the primary industries of agriculture, hunting and forestry, as would be expected in a more rural area. The numbers are still low in absolute terms, although these are areas where self-employment is likely to account for a significant proportion of the total. Manufacturing, substantially skewed towards brewing and distilling, is of a similar scale to primary activities and has grown in recent years, counter to the trend at national level. The services sector accounts for the largest proportion of jobs. Employment in hotels and restaurants in the park (19.4 per cent) is much higher than in Scotland as a whole (5.7 per cent), reflecting the relative importance of the tourism sector. However there are signs of vulnerability in over-dependence on this sector. In addition to issues of seasonality with variations in visitor demand, there are signs of fluctuations in employment figures. The most significant employment losses in the park area between 1998 and 2002 occurred within hotels with a restaurant (minus 120 jobs) and restaurants (minus 94 jobs). The fragility of the tourism industry is reflected in these figures, which show that there were 277 less jobs in the hotels, restaurants and catering sub-sectors. However, there remained 1,278 hotel jobs in the park in December 2002, reflecting the continued importance of tourism to the local economy.

The quality of the natural environment lies at the heart of mainstream tourism but also most other industry within the national park namely forestry, farming and field sports. The area's economy is heavily reliant on tourism in terms of employment and revenue generation with tourism related businesses accounting for 80 per cent of the economy. Specific tourism assets tend to be based on natural and cultural resources; although there are important built visitor attractions, for example the Funicular Railway and the Aviemore Mountain Centre that support outdoor recreational activities such as skiing, water-sports, wildlife/nature watching and walking. Visitor surveys indicate the degree of dependence on natural environmental attractions (see Table 11.1). Continued visitation depends on the quality of these natural assets and visitor spending supports the local economy and contributes to covering the cost of conservation.

Table 11.2 illustrates the spread of revenue generated through tourism and the types of visitor accommodation used. Although there is a large proportion of day

**Table 11.2    Visitor statistics for Cairngorms National Park**

|  | 2004 | 2003 | % change |
|---|---|---|---|
| **Analysis by sector of expenditure (millions)** | | | |
| Accommodation | 35.97 | 34.08 | 6 |
| Food & drink | 23.30 | 22.66 | 3 |
| Recreation | 9.30 | 9.05 | 3 |
| Shopping | 12.93 | 12.62 | 2 |
| Transport | 23.07 | 22.22 | 4 |
| Indirect expenditure | 37.78 | 36.18 | 4 |
| VAT (tax) | 18.30 | 17.61 | 4 |
| Total | 160.64 | 154.42 | 4 |
| **Tourist Days (thousands)** | | | |
| Serviced Accommodation | 1,035.89 | 980.05 | 6 |
| Non-serviced Accommodation | 1,086.25 | 909.45 | 19 |
| Visiting Friends and Relatives | 92.28 | 85.92 | 7 |
| Day Visitors | 850.96 | 894.09 | -5 |
| **Total** | **3,065.39** | **2,869.52** | **7** |
| **Tourist Numbers (thousands)** | | | |
| Serviced Accommodation | 387.39 | 369.02 | 5 |
| Non-serviced Accommodation | 149.75 | 137.92 | 9 |
| Visiting Friends and Relatives | 20.18 | 18.92 | 7 |
| Day Visitors | 850.96 | 894.09 | -5 |
| **Total** | **1,408.28** | **1,419.95** | **-1** |

*Source: Cairngorms National Park Authority 2005 (based on STEAM report 2004)*

visitors as is typical in national parks in the UK, staying visitors balance this out with longer durations of visits.

Visitors to the park are predominately from Scotland, are generally older, with 56 per cent over 45 compared with the Scottish average of 44 per cent. The park attracts a higher than average number of overseas visitors, and compared with the national 'Tourism in Scotland' Survey the visitors appear to be relatively affluent. The visitor profile does vary, however, depending on the reason for visiting. Visitors taking part in more active pursuits tend to be younger, male and are more likely to be from other parts of the UK. Due to the relatively recent designation of park boundaries for The Cairngorms, the CNPA has only been able to estimate visitor numbers. However, a mechanism is now in place which will allow interpretation of existing data to generate figures more specific to the park. This is being undertaken through a tourism economic activity model called STEAM (Scottish Tourism Economic Activity Monitor), developed by Global Tourism Solutions. In 2003 it was estimated that the Park might attract as many as 1.2 million visitors, generating around £240 million a year, but the geographical area used for this research was much greater than the park. At the park level, draft STEAM figures for 2003 gave a total visitor figure of 1.4 million and a total visitor spending of £154 million. Visitor numbers dropped very slightly in 2004, but visitor spend rose to £161 million.

The attitudes and perceptions of respondents to the Cairngorms National Park Visitor Survey were in general positive, with 85 per cent giving their overall visit to the Cairngorms a rating of 8 out of 10 (1 being low and 10 being high). The visitors indicated that there were plenty of things to see and do in the area, and that the park was well managed and cared for. The most appreciated aspects of the Cairngorms area were the beautiful scenery, the mountains and the peaceful easy-going pace of life. There seemed to be no perception of there being too many tourists. There was a high level of awareness (69 per cent) amongst visitors that they were in a national park, and 88 per cent of these were aware of national park status prior to arrival. This did not appear to have been a major influence on their decision to come to the area, however, as only 9 per cent stated that national park status was very important, and 38 per cent said that it was not important at all (see CNPA 2006).

The reasons for visiting the park are many and various, but are largely focused on the wide range of outdoor activities, the natural beauty and the rich cultural heritage which the area has to offer. Walking is recorded as the most popular reason for visiting the area, with beautiful scenery ranked second. There were variations apparent between the different categories of visitors surveyed, including day trippers, short break and long break. Day trippers were more likely to be visiting for walking rather than for general sightseeing or heritage. Visitors on longer breaks were most likely to include general sightseeing and heritage and least likely to take part in active pursuits. The surveys highlighted the perception of wildness and tranquillity as a reason for visiting. The Cairngorms area is seen as offering unspoilt landscapes, inaccessible areas and a feeling of peace and solitude.

The Cairngorms National Park Authority (CNPA) was formally established on 25 March 2003 and took on full operational powers on 1 September 2003. As a statutory Non Departmental Public Body (NDPB) the CNPA is directly funded by the Scottish Ministers. In the longer term, the CNPA objectives will be mainly determined by the National Park Plan, which is the statutory strategic plan for the whole national park area. The first corporate plan outlines the National Park Board's early thinking in both operational and policy terms for the CNPA as an organization, and the long and short term priorities for the park. Four policy themes encapsulate the statutory aims of the park: to foster a park for all, to encourage widespread enjoyment, understanding and appreciation of the special qualities of the area, to develop clear, cohesive strategies for stewardship of the natural resources in the park; and to encourage and support balanced, thriving, stable communities.

The policy theme 'Encourage and support balanced, thriving, stable communities in the park' notes:

> Tourism- the CNP is a large area with tremendous tourism potential. The CNPA aims to establish, working with Tourist Boards and local businesses, a co-ordinated Park-wide approach to sustainable tourism through the preparation and implementation of a sustainable tourism strategy including a marketing strategy and brand for the whole Park, based on the special qualities and attractions of the area and the establishment of a hallmark of quality (Cairngorms National Park Authority 2004, 18).

Tourism is a vital part of the economy in the new park. Local employment relies heavily on tourism and the area benefits from the income that visitors to the area generate. As with all tourist areas however, potential conflict exists between environmental, socio economic and cultural interests and the Cairngorms area is not different. It therefore follows that the resource base on which present and future tourism and tourist activities in the Cairngorms National Park are based must be sustained, since if the resource base is destroyed the tourism that it is based on will surely follow suit.

The development of a Sustainable Tourism Strategy is therefore seen as a priority by the NPA to provide an effective framework for planning, action and evaluation to ensure that tourism is developed in a sustainable and sympathetic manner, whilst ensuring that the heritage and resources are protected. To oversee the development and implementation of this a Sustainable Tourism Officer and Business and Marketing Officer were appointed. Whilst it would appear that sustainable tourism will contribute most to the delivery of the all-important fourth aim 'To promote sustainable economic and social development of the area's communities', the development and management of sustainable tourism has the potential to contribute to all four primary aims for national parks, as set out in the National Parks (Scotland) Act 2000. In addition, Schedule 3 of the Act sets out specific powers for NPAs with respect to tourism, including the provision of information, education and interpretive facilities as well as services to promote the enjoyment of the parks' environment. The Stevens Report (2002) highlighted the fact that this ability to directly provide, manage and intervene, creates interesting opportunities to be proactive, especially in partnership with other stakeholders to provide sustainable tourism products, initiatives and programmes. Furthermore unlike the national parks in England and Wales, the Scottish national parks will be able to encourage people to visit the parks. This power gives the Scottish NPA's the chance to engage in creative destination marketing and to encourage and co-ordinate the marketing of appropriate activities in order to promote sustainable tourism.

The CNPA set up a number of advisory groups and working groups to take forward the imminent work. The Tourism Development Working Group (TDWG) is one of the industry based groups, comprising private sector businesses, the local and national tourism organizations and other relevant public sector organizations involved in tourism in the Cairngorms. The purpose of the group is to identify the priorities for establishing improved co-ordination of tourism related activity the Cairngorms National Park area, and to develop and implement CNP wide initiatives as appropriate. The TDWG recognizes that tourism is all encompassing and in January 2004 identified seven key interlinked issues for tourism in the Cairngorms: the successful co-operation, integration and encouragement of cross-sectoral working for all those involved in tourism in the area; the development of a Sustainable Tourism Strategy and the successful application for, and implementation of, the European Charter for Sustainable Tourism; the development and implementation of a Marketing Strategy and Action Plan for the Cairngorms; the delivery of quality standards that build on nationally recognized standards which are specific to the Cairngorms; the support for and continued development of quality assured products grown, made or available in the area; the need for ongoing research, with easily accessible results, which assesses the needs, opinions and demands of visitors in order to be able to anticipate, meet and exceed visitor expectations; and the enhancement of visitors' experience while in the area, through improved information and interpretation provision.

Further research, analysis and development of actions plans is the next step for the TDWG, whilst recognizing the linkages between the issues and the coordinated approach with partners and other working groups, that must be maintained. The various options (guidelines, tools, codes of conducts) to manage sustainable tourism as outlined in the Stevens Report were considered, and the TDWG felt that the European Charter for Sustainable Tourism provided the most appropriate framework to CNPA at that time. This was particularly the case, as the ECST principles could provide valuable guidance to the relevant National Park Working Groups (access, parks for all, park gateways and information provision to name but a few) in advance of the park strategy being developed and finalized. The ECST in Protected Areas 2002, states that:

> The Charter reflects worldwide and European priorities as expressed in the recommendations of Agenda 21 adopted at the Earth Summit in Rio in 1992 and by the European Union in its 6th Environment Action Programme and Strategy for Sustainable Development. The underlying aims are suited to combining environmental preservation with regional development objectives through tourism: to increase of and support for, Europe's protected areas as a fundamental part of our heritage, that should be preserved for and enjoyed by current and future generations; and to improve the sustainable development and management for tourism in protected areas, which takes into account the needs of the environment, local residents, local businesses and visitors.

The Charter belongs to the Europarc Federation, the umbrella organization of protected areas in Europe and builds on the recommendation of the Europarc study 'Loving Them to Death? – Sustainable Tourism in Europe's Nature and National Parks'. The Charter was one of the priorities defined in the World Conservation Union's action programme for protected areas in Europe, 'Parks for Life' (1994). The Charter outlines a process and provides guidance to ensure that the park authority is able to manage sustainable tourism effectively and innovatively through 12 principles including: protecting and enhancing the natural and cultural heritage; understanding and meeting visitor needs and ensuring quality; communicating the special qualities of the area; encouraging tourism products relating to the protected area; training relating to the protected area and sustainable tourism; maintaining the local quality of life; increasing benefits to the local economy; and monitoring and influencing visitor flows. It provides a framework specifically for protected areas to create a structure and context for working in partnership on sustainable tourism so that a sustainable tourism strategy can be developed which includes a five year action plan. It also places a commitment on the park authority to undertake on-going consultation with the private sector; and to devise a means by which targets are set and progress is evaluated.

For the Cairngorms the 'checklist' aspect is particularly significant as it has helped to avoid missing key areas or actions that should be covered in the early stages of strategy development. The Charter process also involves an objective assessment of the strengths, weaknesses, opportunities and threats facing the Park and requires a review and evaluation mechanism. The TDWG recognizes that the process contained within this framework is as important, if not more so, than gaining the Charter itself. To be successful, it is important that all the key stakeholders are involved in its development and implementation. The local private sector have been actively involved from the start as ultimately they are the key beneficiary, though individual businesses are unlikely to see the full benefits until later in the process. The parks strategic objectives reflect this broader, bottom-up perspective to protected area management with almost equal emphasis placed on: working together; enterprise and economy; environmental management and conservation; visitor management; community involvement.

## Discussion

Although in the early stages of implementation, the innovative models developed in Scotland are worth examining in the search for models that blend protected area and regional development policies. Scotland's approach to national parks is firmly based on the principles of sustainable development, attempting to integrate the needs of local communities living and working in the park with the needs of heritage conservation and enhancement. The model seeks to avoid conflicts and reconcile competing interests through partnership working. The approach differs from the dominant international approach to national parks which is based on the primacy of protecting the natural environment. It could be argued that this is a modern model to reflect contemporary circumstances in a post industrial country attempting to address taxing social and economic requirements. The robustness of this model will be tested in its ability to reconcile tensions between environmental protection and sustainable development for the benefit of socially excluded communities within park boundaries. The influence of tourism, in particular NBT, is crucial but a lack of clear, measurable indicators of change hinders rational remedial action. The connections between tourism and the environment are acknowledged by policy-makers but only in a general sense. One test of the national park model for Scotland will be its ability to ensure that revenues from NBT benefit local communities and are used to offset costs of management/conservation policies. To assist in this, marketing of NBT must be integrated into National Park policies to sensitively combine attracting lucrative niche markets with encouraging social inclusion through local recreation.

The Cairngorms as a case study area epitomizes a 'living landscape' that incorporates a rich mix of natural and cultural heritage resources yet recognizes the full range of social and economic problems associated with European peripheral rural regions. The Cairngorms are now a designated area that has clear regional development needs, where protecting the natural environment is viewed as a direct means to grow an economy primarily through catering for tourist needs. Innovative approaches are illustrated through the adoption of management mechanisms with international recognition (ECST) to facilitate tourism development in a sustainable way through involvement of tourism businesses from the outset TGWD. It also exemplifies a protected area that is based on and maintains a high degree of local community participation and control. It is too early and there is insufficient evidence to measure success of this new model but initial indicators tend to be positive with adjacent areas, such as the Angus Glens currently looking to become part of the national park to share in its perceived economic development advantages. Nevertheless more mature, established protected areas in Europe should certainly consider these policies as modern innovative approach that addresses many long standing intractable problems associated with the economics of protected areas and their rural hinterlands.

## References

Bishop, K., Phillips, A. and Warren, L.M. (1997), Protected Areas for the Future: Models from the Past, *Journal of Environmental Planning and Management*, 40:1, 81–110.

Boyd, S. W. (2000), 'Tourism, national parks and sustainability', in Butler, R.W. and Boyd, S.W. (eds), *Tourism and National Parks: Issues and Implications*. (Chichester, John Wiley & Sons Ltd), 161–86.

Boyd, S.W. and Butler, R.W. (2000), 'Tourism and national parks: the origin of the concept', in Butler, R.W. and Boyd, S.W. (eds), 13–27.

Buckley, R. (2003), 'Pay to Play in Parks: An Australian Policy Perspective on Visitor Fees in Public Protected Areas', *Journal of Sustainable Tourism* 11: 1, 56–73.

Butler, R.W. and Boyd, S.W. (eds), *Tourism and National Parks: Issues and Implications*. (Chichester, John Wiley & Sons Ltd).

Cairngorms National Park Authority (2004), *Cairngorms National Authority: Corporate Plan 2004-2007*. (Grantown-on-Spey).

Cairngorms National Park Authority (2005a), *Tourism and the Cairngorms National Park: introducing the park's sustainable tourism strategy and survey of visitors*. (Grantown-on-Spey).

Cairngorms National Park Authority (2005b), *A Strategy and Action Plan for Sustainable Tourism*, CNPA. (Grantown-on-Spey).

Cairngorms National Park Authority (2006), *State of the Park Report 2006*, CNPA. (Grantown-on-Spey).

Eagles, P.F.J. (2002), 'Trends in Park Tourism: Economics, Finance and Management', *Journal of Sustainable Tourism* 10:2, 132–53.

Edwards, T., Pennington, N. and Starrett, M. (1993), 'The Scottish Parks System: A Strategy for Conservation and Enjoyment', in Fladmark, J.M. (ed.), *Heritage*. (London, Donhead Publishing), 141–51.

Hardy, A., Beeton, R.J.S., and Pearson, L. (2002), 'Sustainable Tourism: An Overview of the Concept and its Position in Relation to Conceptualisations of Tourism', *Journal of Sustainable Tourism* 10:6, 475–94.

International Union for Conservation of Nature and Natural Resources (IUCN) (1994), *Guidelines for Protected Area Management Categories*. (Switzerland, Gland).

Jamal, T. and Eyre, M. (2003), 'Legitimation Struggles in National Park Spaces: The Banff Bow Valley Round Table', *Journal of Environmental Planning and Management* 46:3, 417–41.

Laarman, J.G. and Gregersen, H.M. (1996), 'Pricing policy in nature-based tourism', *Tourism Management* 17:4, 247–54.

Leitmann, J. (1998), 'Options for Managing Protected Areas: Lessons from International Experience', *Journal of Environmental Planning and Management* 41:1, 129–43.

Liston-Heyes, C. and Heyes, A. (1999), 'Recreational benefits from the Dartmoor National Park', *Journal of Environmental Management* 55:2, 69–80.

Lovelock, B. (2002), 'Why It's Good to be Bad: The Role of Conflict in Contributing Towards Sustainable Tourism in Protected Areas', *Journal of Sustainable Tourism* 10:1, 5–30.

McCarthy, J., Lloyd, G. and Illsley, B. (2002), 'National Parks in Scotland: Balancing Environment and Economy', *European Planning Studies* 10:5, 665–70.

Moir, J. (1997), 'The designation of valued landscapes in Scotland', in MacDonald, R. and Thomas, H. (eds) *Nationality and Planning in Scotland and Wales*. (Cardiff, University of Wales Press), 203–242.

Parker, G. and Ravenscroft, N. (1999), 'Benevolence, nationalism and hegemony: fifty years of the National Parks and Access to the Countryside Act 1949', *Leisure Studies* 18:4, 297–313.

Scottish Executive (2000), *A New Strategy for Scottish Tourism*. (Edinburgh, The Stationary Office).

Scottish Executive (2002), *Tourism Framework for Action: 2002 – 2005*. (Edinburgh, The Stationary Office).

Scottish Executive (2006) *Scottish Tourism: the next decade – a tourism framework for change*. (Edinburgh, The Stationary Office).

Scottish Natural Heritage (1999), *National Parks for Scotland: Advice to Government*, Scottish Natural Heritage. (Edinburgh).

Stevens, T. (2002), *Sustainable Tourism in National Parks and Protected Areas: An Overview*, Scottish Natural Heritage. (Edinburgh).

The Tourism Company (2005), *Cairngorms National Park: a Strategy and Action Plan for Sustainable Tourism*, Cairngorms National Park Authority. (Grantown-on-Spey).

Van Sickle, K. and Eagles, F.J. (1998), 'Budgets, pricing policies and user fees in Canadian park's tourism', *Tourism Management* 19:3, 225–35.

Warren, C. (2002), *Managing Scotland's Environment*. (Edinburgh, EUP).

World Tourism Organization (WTO) and United Nations Environment Programme (UNEP) (1992), *Guidelines: Development of National Parks and Protected Areas for Tourism*. (Madrid, WTO).

**Internet Links**

HMSO (2000), <http://www.scotlandlegislation.hmso.gov.uk/legislation/scotland/acts2000/20000010.htm>.

Scottish Office (1997), 'Planning Bulletin, issue sixteenth December, 1997', *National Parks for Scotland*; <http://www.scotland.gov.uk/library/documents1/dd-pl16a.htm>.

Chapter 12

# Protected Areas and Regional Development Issues in Northern Peripheries: Nature Protection, Traditional Economies and Tourism in the Urho Kekkonen National Park, Finland

Jarkko Saarinen

## Introduction

Natural areas in the North have increasingly been objects of economic, social and political interest in recent years. In the European context, the North as a region has become prominent when defining European Union regional and economic policies, nature protection and the use of natural and energy resources, climate change and environmental issues in general, and also the position of indigenous peoples.

Economically, northern areas have represented peripheries that are open to the utilization of their resources, such as forests, minerals and fish stocks, by the centres (cores). Northern nature has also been used by the local people for their traditional forms of livelihood, such as hunting, fishing and reindeer herding. On the other hand, those parts of northern peripheries which have lacked any 'real' economic importance and have thus remained as relatively amenity-rich landscapes have been some of the first targets of institutional nature conservation. These 'white areas' on the economic maps have increasingly become the focus of new regional development discussions in recent times, however, and have been subordinated to the needs of various global-scale economic utilization projects. Plans to open protected wilderness areas in Alaska to oil and gas production exemplify this kind of process in a North American context. A similar global-scale factor influencing the natural areas in the whole circumpolar region has been the increasing interest shown by the tourism industry, and especially new forms such as nature-based tourism and ecotourism.

Recently tourism has been widely promoted and used in peripheral natural areas as a replacement economy for traditional means of livelihood. The future prospects opened up by tourism in particular are seen as a potential instrument for controlling

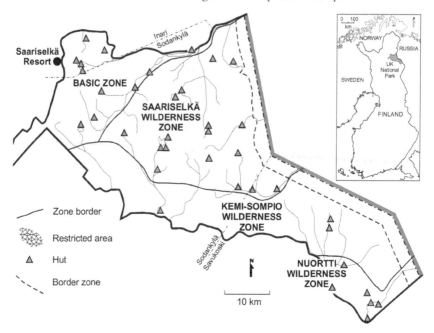

**Figure 12.1   Location and management zones of the UK National Park**
*Source: Drafted by author*

the economic transition and its social consequences in peripheral communities and regions. In addition, tourism is regarded as one major activity that can be practiced and developed in a close mutual relationship with existing protected areas. In many cases the tourism development potential can also be used as an argument for increasing nature protection. Tourism has been understood as a road to peripheral and community development for a long time (see Telfer 2002), but since the early 1990s it has also represented a potential tool for putting sustainable development into practice on a local scale (see Berry and Ladkin 1997, Saarinen 2006).

In Finland, too, and especially in northern Finland, the tourism industry has become an important, growing and crucial element in regional economies and in everyday life. Tourism and tourists provide more employment opportunities in Finnish Lapland, for example, than any other field of the economy that makes direct or indirect use of natural resources, with an estimated employment effect of over 4,000 full-time jobs in 2,000 (see Saarinen 2003). The growing economic role of tourism has also made it a social and political issue, and tourism has been increasingly used as a medium for achieving many socio-cultural and economic goals on a regional or local level which may conflict with the traditional and modern economies and uses of northern natural resources, including nature protection and forestry.

The main purpose of this chapter is to discuss the interplay between protected areas and regional development, using the Urho Kekkonen (UK) National Park as a case-study area. Following an introduction to the historical background and the present management situation of the UK National Park, the relationship between the national park, regional development issues and tourism will be discussed.

## Purpose and Nature of the UK National Park

The Urho Kekkonen National Park is located in the municipalities of Inari, Savukoski and Sodankylä in north-eastern Finland (see Figure 12.1). It was established in 1983 and is managed by Metsähallitus (the Finnish Forest Service). This is a state-owned enterprise (or 'semi-private business') whose main tasks are forestry, that is to supply wood to the forest industry, and the management of most of Finland's protected areas. The funding for its nature protection work is channelled from the state budget, and part of the forestry revenue is also allocated for conservation purposes.

The UK National Park was established to protect forest, mire and fell habitats in forested parts of Finnish Lapland and to preserve the preconditions necessary for practicing reindeer herding and for traditional hiking activities in a wilderness environment. The UK National Park is the second largest among the present 35 national parks in Finland, with an area of 2,550 km², and is one of the most heavily visited by tourists, with approximately 165,000 visits in 2005 (see Käyntimäärät ... 2005, Saarinen 2005).

The UK National Park represents the northern taiga forest and is characterized by a combination of fell and mire landscapes. The northern part is a rugged wilderness characterized by ravines and steep slopes, while the south-western part features extensive open bogs. In the south there is a typical forest wilderness area with isolated fells, pine forests and mossy spruce forests. The larger mammal species inhabiting the park include the brown bear (*Ursus arctos*), lynx (*Lynx lynx*), elk (*Alces alces*) and otter (*Lutra lutra*), while wolves (*Canis lupus*) and wolverines (*Gulo gulo*) regularly visit the area (see Urho Kekkonen ... 2001).

Reindeer herding, hunting and fishing have long traditions in the area, and there are also related facilities and objects belonging to the cultural heritage, such as reindeer fences, herders' huts and restored Skolt Sámi settlements. Reindeer herding has a special status in the national park and represents a major source of livelihood over a wider area. The maximum number of reindeer allowed in the park area is 20,000. Local residents of the municipalities of Inari, Savukoski and Sodankylä are allowed to hunt in the park under the conditions of the Finnish Hunting Act, and visitors may fish in certain parts of the park if they obtain a Metsähallitus recreational fishing permit. The picking of berries and mushrooms is permitted except in areas that would otherwise be restricted in this respect (see Urho Kekkonen ... 2001). There is also some small-scale gold digging practised in the park area, with 21 claims in force in 1998. Gold panning or digging is regulated by the Mining Act, which takes precedence over the Conservation Act. Thus small-scale gold panning without large motorized machines can be carried on inside national parks.

The UK Park has two visitor centres, in Sodankylä and Savukoski, and an information centre at the tourist resort of Saariselkä in Inari. There are five huts in the park for daytime use and altogether 38 freely available or bookable huts for over-night use by visitors. The national park is divided administratively into four management zones (see Figure 12.1), each with its own specific rules (see Urho Kekkosen ... 2001). The basic zone is in general more closely regulated, better developed and more crowded than the three wilderness zones. It has a dense network of signposted hiking trails, and visitors are encouraged to use only these trails and other recreational facilities provided, such as fireplaces. There are also designated trails for mountain biking, and about 200 kilometres of maintained skiing tracks in the winter season. The wilderness zones are relatively undeveloped in terms of recreational facilities and trails. Visitors are basically free to make a fire and pitch a tent wherever they wish in the most remote Kemi-Sompio Wilderness Zone, but they are encouraged to use specific sites for this in the Saariselkä and Nuortti Wilderness Zones. In addition, there are separate restricted zones in the southern part of the park to protect nesting birds. These areas are also open for hiking and the picking of berries outside the nesting season, but they are rather inaccessible in summer because of the wet mire terrain. The restored historical Skolt Sámi settlement areas and the border zone between Finland and Russia are also restricted zones.

## Conflicting Interests in Europe's 'Last Remaining Wilderness'

The present Urho Kekkonen National Park is part of the larger Koilliskaira area, a region in north-eastern Finland characterized by old forests and wilderness environments that was effectively introduced to the general public in the late 1950s and early 1960s by the novelist Kullervo Kemppinen, whose books *Lumikuru* (The Snow Gully 1958) and *Poropolku kutsuu* (The Call of the Reindeer Path 1962) painted a sublime picture of the wilderness character of the present area of the UK National Park, its special features and the hikers that visit it. At that time the area of the present national park was known as Saariselkä, based on the Sámi name Suolocielgi ('a ridge of islands' which probably refers to the fells, which resemble islands in an otherwise flat mire terrain). The region consequently became known as a hikers' paradise, and the effect of these books was seen directly in the numbers of tourists. While the average annual increase in the numbers of visitors in the 1950s was well under 40 percent, the number doubled in the year following the publication of *Lumikuru*, and this vigorous expansion continued into the early 1960s (see Saastamoinen 1982). This also resulted in some conflicts between the recreational users. There were some reports of conflicts between 'hikers' and 'tourists' in the present park area as early as the 1970s, for example, when differences became apparent in attitudes and practices towards the use of nature in wilderness areas (see Saarinen 1998a).

At the same time, however, the increased timber needs of the wood-processing industries in the North forced Metsähallitus to plan forest fellings in the Saariselkä area, for which purpose a dense network of forest roads was planned. In response to this, the Finnish Tourist Association, the hiking and skiing organization *Suomen Latu* and the Finnish Nature Conservation Association made a proposal in 1961 that the area should be protected from all felling and preserved as a roadless tract of wilderness. The forestry plans large clear cuttings, which were typical of the time and reflected the increasing needs for industrial timber and the fact that many other previously forested areas in southern Finnish Lapland had already been felled. It should be noted that there were no real means of importing wood in the 1960s and 1970s, and the industrial needs of Finnish forestry sector would have had to be satisfied mainly by reducing domestic forest resources. Thus the conflict situation was likely to increase in subsequent years.

Later, in 1967, the Finnish Nature Conservation Association proposed that the whole of Koilliskaira should be protected entirely, comprising an uninterrupted area of some 5,000 km². Other corresponding proposals and demands were made, but it took years before these led to any concrete results, despite heated, value-loaded discussions on the issues of forestry, conservation and tourism (see Häyrinen 1979; Saarinen 1998b). The resulting publicity in the media raised Koilliskaira to a position of national significance as one of the last extensive wilderness areas in Europe (see Häyrinen 1989; Borg 1992).

Two committees in the 1970s, the Koilliskaira Committee set up by Metsähallitus in 1972 and the Parliamentary National Park Committee in 1976, made suggestions on how to proceed with possible nature protection in the region. The latter committee, for example, stated that the region should be protected as a national park comprising an area of 3,200 km². The subsequent parliamentary decision of 1978 to develop the network of protected areas in Finland did not include the Koilliskaira region, however (see Urho Kekkosen... 2001).

Following a long dispute, the area was finally protected in 1980 under the name of the Urho Kekkonen National Park (2,550 km²). The final phase was clearly a political process, or one might even say a manoeuvre. The conservationists and related actors introduced the idea that the area should be protected by the time of the 80th birthday of the Finnish President, Urho Kaleva Kekkonen (1900–1986), who had visited the area several times, and suggested that the planned national park should be the 'nation's gift' to this highly respected and powerful figure (see Urho Kekkosen ... 2001, 40). This was politically very difficult to oppose or even challenge at the time, and in practice the law to this effect came into force in 1983.

## Management Issues in the UK National Park

*The Idea of 'Wilderness'*

As traditional forms of livelihood, especially reindeer herding, fishing and local hunting, were still being practised in the areas the status of the UK National Park in terms of the World Conservation Union (IUCN) categories has been relatively 'low'. Instead of the present Category II, that of a national park, it was placed until 2003 in Category IV, a habitat and species management area set aside mainly for conservation through management intervention (see IUCN 1994).

This lower status category did not necessarily reflect the nature and situation of the relatively large, diverse area, but rather the challenges and problems entailed in introducing western and mainly Anglo-American ideas of nature protection in other contexts, in this case in Finland and Finnish Lapland. The differences in the idea of what constitutes a wilderness, Category Ib in the IUCN list, illustrate this problem very well. In the Finnish context, a wilderness (*erämaa*) has historically implied a region with no permanent community settlement that is of economic and cultural importance in terms of hunting and fishing (see Lehtinen 1991). As such, wilderness areas were considered integral parts of the medieval system of society, with its hunting economy, in which the hunting areas, and especially their taxation, were often assigned to families or communities (see Hallikainen 1998). In spite of its nature as a legal 'relic', the traditional concept of wilderness has for a reasonably long time been the dominant idea in Finland, and it continues to be so in many contexts and in peripheral places such as northern Finland and Finnish Lapland (see Saarinen 2005).

The traditional idea of wilderness is manifested in the current Finnish Wilderness Act (see Erämaalaki 1991), for instance, according to which wilderness areas can be designated in order to:

1. Maintain their wilderness character
2. Protect the sámi culture and traditional means of livelihood or
3. Develop the diversified use of nature and possibilities for different (economic) uses.

Apart from the first, somewhat vague, open definition, the goals of the act are notably similar to the concept of wilderness in the hunting culture. The traditional wilderness as described in the act is a resource contained with the culture and the economy, and it is defined mainly through its human use and local value (see Saarinen 2005). In contrast to this, the western idea of a protected wilderness is in many ways based on a historical perspective in which a natural or wilderness area and any primitive inhabitants and cultures that it contained were seen as the antithesis of, and even a hindrance to, civilization and development, especially in Anglo-American societies. The western concept of wilderness or nature was thus mainly formed through conquering and transforming wilderness areas, and through the juxtaposition of nature and culture – not through living in and from the wilderness (see Short 1991, 5–10). As Roderick Nash (1982: xiii) states in his book *Wilderness and the American Mind*, 'Civilization created wilderness'. In saying this he is referring to the clear line drawn between wilderness and organized society.

In North America, the increasing elimination, or 'civilization', of the wilderness finally resulted in a need to protect the remaining areas. As a consequence, the world's first Wilderness Act was passed in the United States in 1964 (see Public Law 1964, also Hall 1998). This places wilderness areas outside society and culture:

> A wilderness, in contrast to those areas where man and his own works dominate the landscape, is hereby recognized as an area where the earth and its community of life are untrammelled by man, where man himself is a visitor who doesn't remain [...].

This idea is somewhat opposed to the traditional concept of wilderness that formed the basis of the Finnish Wilderness Act, a contradiction which was probably reflected in the past status of the UK National Park in the IUCN list.

*The Changing Management Context*

The UK National Park has a board of representatives of national, regional and local interest groups which defines and supports the management goals. The management strategy is processed according to the principles of participatory planning, and the national park works closely with local communities that are dependent on reindeer herding and also consults the Sámi Parliament on larger management issues and practices that may affect the traditional livelihood and Sámi culture of the region (see Urho Kekkosen ... 2001).

Major parts of the UK National Park lie within the Home Region of the Sámi, who are the only indigenous people inside the present European Union, forming an ethnic minority in Finland, Norway, Sweden and also north-western Russia (the Kola Peninsula). Approximately 4,000 out of about 6,400 Sámi in Finland live in this Home Region, which is currently owned and managed by the state (see Seurajärvi-Kari et al. 1995). This state ownership, and especially the management of land, is contested by the Sámi, and their local rights and needs concerning land use have aroused conflicts between reindeer herding and forestry. Such conflicts do not have a direct link to the management of the UK National Park, because no forestry is practiced there, but they do form a wider political management context and there are also minor issues between reindeer herding and the growing nature-based tourism which are evident in the park area as well.

The conflict between reindeer herding and nature-based tourism is mainly related to the use of winter pastures. The winter time is crucial for the survival of the reindeer, and tourism infrastructures and tourists themselves may limit the use of pastures, which can lead to the over-grazing in other areas and losses in meat production. There are also discussions concerning the harassment of calving reindeer in spring by the increasing numbers of cross-country skiing tourists, which may lead to the loss of newborn calves (see Helle and Särkelä 1993). Based on a comparative study (between 1988 and 2000), it seems that growing touristic uses of the national park's western areas (the basic zone), where most of the tourism infrastructure is located, have probably resulted in a higher reindeer density, and thus over-grazing, in the distant Saariselkä Wilderness Zone (see Niva 2003). It is worth noting that the ecological problem of over-grazing and erosion in the wilderness zone and socio-economic problems of declining meat production in local communities that are dependent on reindeer herding are indirect results of the increased touristic use of the park's basic zone.

The UK National Park is part of the European NATURA 2000 Network, together with the adjacent Kemihaara Wilderness Area, Sompio Strict Nature Reserve and Uura-Aava, Vaara-Aava and Nalka-Aava Mire Protection Areas. The national park has not been widely involved in EU LIFE projects that could be used to support habitat and species protection, but its administration is increasingly becoming involved in international collaboration, especially with nature protection areas on the Kola Peninsula of north-western Russia (see Urho Kekkosen ... 2001). In addition, the UK National Park is one of the main tourist attractions in north-eastern Finland, and especially in the municipalities in which it is located. Thus it has been involved in local regional development projects supported mainly by EU funding instruments (for example ESR, LEADER). In this respect the local situation has changed dramatically from that prevailing in the 1980s. The national park was initially opposed by the local municipalities and also by entrepreneurs, and even after its establishment there were some local regional development actors who believed that its protection could be overruled in order to utilize the area's forest and potential mining resources (see Saarinen 1998b). The delay in obtaining state compensation for its conservation aroused particularly vehement criticism in the 1980s, but attitudes changed during the following decade (see Borg 1992).

There are several possible reasons for this change in attitude, which also implies an ongoing change in the management context. First, the project-driven EU regional policy as increasingly come to view the park and its larger management organization as 'good partners' who possess administrative abilities, knowledge and other human capital and skills in a region that is otherwise characterized by shrinking and ageing population structures and the out-migration of young and educated people. It is general knowledge that EU funding applications and tools with intensive reporting duties and other forms of bureaucracy are not always simple to administrate in the peripheries. Second, the growing trend for nature-based tourism and ecotourism has integrated the protected areas, especially the national parks, which have recreational purposes, more deeply into the regional (tourism) economy. National parks with nature-based tourism activities are increasingly being seen as tools for regional development. As a result, many municipalities and other regional development actors have actively proposed additional national parks in northern Finland, and some of these actors are the same ones who opposed national parks in the 1970–80s, for example! Third, and partly related to the previous issues, there seems to be a real or perceived lack of alternatives to tourism in peripheral areas and communities, which has almost led to an absolute necessity to develop tourism activities and resources. In amenity-rich peripheries tourism activities are deeply integrated into the use of natural areas (see Hall and Boyd 2005), and national parks, with a positive international and national status and image, recreational infrastructure and often relatively good accessibility, potentially provide good platforms for this.

In spite of its growing economic role, the establishment of the UK National Park failed to halt the polemic over the economic exploitation of the region entirely in 1990s. A previous example of this is the disagreement that emerged over the costs of protecting the last old virgin forests in Finland in the mid-1990s. It was claimed in one major survey, for example, that much more profit could be obtained from felling the forests of the present UK National Park area than could be made from nature protection and tourism (Pohjois-Suomen ja Pohjois-Karjalan ... 1995). It was calculated that the total yield per hectare that would accrue from the economic exploitation of the old forests for which protection orders were pending would be three times greater than that achieved through protection and related tourism use, and that the employment effects of exploitation would likewise exceed those of the latter policy three-fold. These calculations were nevertheless based on an exploitation period of 25 years, after which no real income would be obtained from forestry for the next 150–170 years, given the latitude and climatic conditions of the UK National Park. Protection with growing nature-based tourism should be beneficial throughout that time, so that this will most probably be more profitable in terms of income and jobs in the long run (see Saarinen 1998b, Power 1996).

Whatever the situation may be in the calculations and in reality, the above events related to the history of the establishment of the UK National Park and the debate over the costs of protection were characterized by the fact that tourism and nature

**Table 12.1    Numbers of visitors to selected national parks in northern Finland in 1995, 2000 and 2005**

| *National park* (area size in km$^2$) | 1995 | 2000 | 2005 |
|---|---|---|---|
| Lemmenjoki (2,850 km$^2$) | 10,000 | 10,000 | 10,000 |
| Oulanka (270 km$^2$) | 100,000 | 145,000 | 173,500 |
| Riisitunturi (77 km$^2$) | 2,000 | 10,000 | 7,000 |
| Urho Kekkonen (2,550 km$^2$) | 200,000 | 150,000 | 165,000 |

*The figures for 1995 represent estimates, while the later figures are based on more advanced calculations but still represent estimates for the total number of visitors.*
*Source: Metsähallitus*

protection were linked closely together to form one alternative to forestry. It would seem in the light of this example that a certain synergy exists between use of the natural environment for tourism and its protection (see Budowski 1977). In the case of the UK National Park no significant contradiction was perceived between nature protection and the development of tourism, at least at the initial stage, perhaps even the reverse (see Borg 1992).

In response to the changing management and regional development contexts, the UK National Park and Metsähallitus are evolving into potential actors in regional development processes and are functioning as intermediaries between local and international interests in both nature protection and regional development issues. In this respect, and especially from the regional development perspective, the role of tourism has become a crucial one in the national parks and other protected areas of peripheries.

**The Growing Role of Tourism**

Tourism is perhaps the latest significant form of economy and consumption to use the natural areas in the north, and internationally it is considered highly important and a form of economy that makes great use of natural environments in its operations (see Hall and Jenkins 1998). Large amounts of public money are being spent nowadays on the development of tourism in the various peripheral regions of the European Union, as this is the case in the northern parts of Finland, too.

Tourism has been intensively developed in northern Finland in the past few years, so that statistics indicated a total of over 3.6 million overnight visits in 2004, including about 1.8 million person/nights spent in tourist accommodation in Lapland (Tourism Statistics ... 2005), where most of the larger national parks are located. The total average sum of visitors to the Finnish national parks more than doubled in the 1990s (see Saarinen 2005), but this development was not equally distributed among them, as the national parks which are relatively accessible received larger numbers of visitors than those which are more remote from the main tourist destinations or roads. Trends in visits to those national parks in northern Finland that have uniform follow-up statistics extending over several years are shown in Table 12.1.

The UK National Park thus represents a relatively intensively used area among the national parks in northern Finland. Most tourists visit only the basic zone (over

150,000 visits annually), as this is easily accessible from the Saariselkä tourist destination, which has a capacity of approximately 12,000 bed places at the moment. Only a minority of the visits take place in the wilderness zones, the estimate for all three combined being 10,000 backpackers, thus accounting for about 60,000 nights spent in the park annually (see Saarinen 1998a).

The Saariselkä tourist destination, to which the UK National Park is closely linked, is located alongside a major tourist route (E4) to the North Cape in Norway and is served by an airport which is located about 27 kilometres further north. The resort represents a large tourism business district with a variety of services, including restaurants, nightclubs, pizzerias, a brewery, shops, a spa, and so on. Although tourism in Saariselkä today and the attractiveness of the region as a whole are largely an outcome of the founding of the national park and protection of the wilderness character of the environment, there are a wealth of ideas surrounding tourism and its development.

There are no accurate or up-to-date statistics on the turnover for tourism in Saariselkä or any figures on the economic importance of the UK National Park in the field of nature-based tourism. The annual total turnover of the tourism industry in the area in 1998 was about 25 million euros, and the industry employed over 200 people living in the municipalities of Inari, Savukoski and Sodankylä (see Alakiuttu and Juntheikki 1999; Saarinen 1998b). The proportion of this account for by nature-based activities (recreation and nature safari services) was about 7 percent, which still represents a relatively low figure compared with many other tourist destinations in northern Finland (see Saarinen 2003). This means that there is still a large unused potential for nature-based tourism that could be utilized for the benefit of tourism and regional development, although more effective utilization might also raise problematic issues with respect to nature protection in the areas in the future.

Tourism in Saariselkä resort area has expanded rapidly in terms of both the number of visitors and the construction of an infrastructure of tourist services since the early days of discussions regarding nature protection in the area. Where the accommodation capacity in 1960 was about 200 beds (see Saastamoinen 1982; Saarinen 1998a) and it has increased to some 12,000 beds by 2005. A recent general tourism plan for the region allows for a further increase in capacity to some 20,000 beds (see Saariselän yleiskaava ... 1993) and for the provision of services that may in future include sports and motor sports arenas, gold panning, husky and reindeer sledge routes and increased numbers of hiking, skiing and snowmobile routes (see Saariselän liikunta-ja virkistyspalvelut ... 1995). All this will most probably mean major changes in both the physical environment and the images and motives associated with tourism in Saariselkä, and also in the closely related UK National Park. This raises the question of whether the latter can still maintain its image as 'the last wilderness in Europe' which was created by the media in the light of the conservation debate in the 1960s and 1970s and is still intensively used in the advertising of Saariselkä.

The annual growth in nature-based tourism in Finland is estimated to be as high as 8–10 per cent (see Ympäristöministeriö 2002). Generally speaking, this may be an over-estimation which will not easily be reached at most nature-based tourism destinations. Even lower rate of direct growth in the touristic use made of natural areas and in new forms of tourism activity could still significantly influence the use and character of natural areas such as the UK National Park, however, and also affect the relationships between different forms of use, including local traditional means of livelihood. Indeed, structural changes related to tourism have been visibly present in the development of tourism in the UK National Park and the Saariselkä tourist destination during the past decade. Traditional activities have been backpacking, hiking and fishing in the summer season and skiing in the winter (see Kauppi 1996; Saastamoinen 1982). Snowmobile trekking has become one of the most central and most visible forms of new nature-based tourism activity outside the UK National Park, but traditional nordic skiing is still the most important form inside the park during the winter season. Other activities that are gaining in popularity are horse safaris, mountain biking, canoeing/kayaking and dog sledge safaris.

## Conclusions

Peripheral protected areas are facing increasing economic interest and pressures from regional development actors. Tourism especially has been widely promoted in recent years and has been used in peripheral areas as a replacement for the traditional forms of economy. The future prospects for nature-based tourism are encouraging, and it also represents a potential mechanism for utilizing protected areas for the benefit of regional development. Alongside these trends, the management contexts of the national parks are changing and the parks are becoming integrated more deeply into local and regional socio-economic contexts and plans. As a result, the Urho Kekkonen National Park has the potential for becoming an important actor in regional development processes and is increasingly coming to represent an intermediary between local and international interests in both nature protection and regional development issues.

In the case of the UK National Park tourism offered a basis for arguments which would have been difficult to formulate on nature protection grounds alone, providing an opportunity to point to the economic benefits of protection in the form of jobs and incomes. This also highlighted the economic role of protected areas in peripheries. Nevertheless, nature protection together with increasing tourism has not completely replaced the traditional meaning structures, uses and values of the park or of northern nature in general.

The future challenge for the amenity-rich peripheries will lie in the relationship between nature-based tourism development in protected areas and regional development. Nature-based tourism and tourist activities are potentially good tools for regional development, the production of well-being and the sustainable use of the environment and resources in peripheries. Tourism is primarily an economic industry, however, which often has its own goals that may be contradictory to the regional development and nature protection objectives adopted in specific places. In the case of the UK National Park, and more generally, tourism can represent a tool that can be used for nature protection and regional development purposes, but it should be used with care in the future in order to avoid a situation in which a means become an end. In this respect natural environments such as national parks and peripheral communities that rely on local uses and meanings of natural resources are in a critical position. Without integration into the management goals of protected areas, local communities and an active purpose to create mutual benefits along with these localities, the increasingly globalized tourism industry may even create uneven and unsustainable regional development outcomes in peripheries.

## References

Alakiuttu, K. and Juntheikki, R. (1999), 'Matkailun aluetaloudelliset vaikutukset Inarin kunnassa', in K. Alakiuttu, R. Juntheikki, J. Saarinen and P. Kauppila (eds), *Inarin kunnan matkailututkimus*, Nordia Tiedonantoja 4: 3–43.

Berry, S. and Ladkin, A. (1997), 'Sustainable tourism: a regional perspective', *Tourism Management*, 8: 433–40.

Borg, P. (1992), *Ihmisten iloksi ja hyödyksi: Vastuun luonnonsuojelupolitiikkaa rakennemuutos-Suomessa*. Suomen Luonnonsuojelun Tuki Oy, Helsinki.

Budowski, G. (1977), 'Tourism and conservation: conflict, coexistence, or symbiosis', *Parks*, 3–7.

Erämaalaki (1991), Suomen säädöskokoelma 1991, N:o 62, 129–131.

Hall, C.M. (1998), 'Historical antecedents of sustainable development and ecotourism: new labels on old bottles?' in C.M. Hall and A.A. Lew (eds), *Sustainable tourism: geographical perspective*, (Longman, New York), 13–24.

Hall, C.M. and Boyd, S. (2005), 'Nature-based Tourism in Peropheral Areas: Introduction', in C.M Hall and S. Boyd (eds), *Nature-based Tourism in Peripheral Areas: Development or Disaster?*, (Channelview Publications, Clevedon), 3–17.

Hall, C.M. and Jenkins, J.M. (eds) (1998), *Tourism and recreation in rural areas*, (John Wiley, Chichester).

Hallikainen, V. (1998), 'The Finnish wilderness experience', *Metsäntutkimuslaitoksen tiedonantoja*, 711, 1–288.

Helle, T. and Särkelä, M. (1993), 'The effects of outdoor recreation on range use by semi-domesticated reindeer', *Scandinavian Journal of Forest Research*, 8, 123–133.

Häyrinen, U. (1979), *Salomaa*, (Kirjayhtymä, Helsinki).

Häyrinen, U. (1989), *Koilliskaira: Urho Kekkosen kansallispuisto*, Otava, Helsinki.

IUCN (1994), *Guidelines for Protected Areas Management Categories*. (IUCN, Cambridge and Gland).

Kauppi, M. (1996), 'Suomen luonto kansainvälisenä matkailutuotteena', *Suomen Matkailun Kehitys Oy:n julkaisuja* A: 70: 1–39.

Käyntimäärät kansallispuistoittain 2005 (2005), www.metsahallitus.fi (10.2.2006).

Lehtinen, A. (1991), 'Northern natures: a study of the forest question emerging within the timber-line conflict in Finland', *Fennia*, 169: 57–169.

Nash, R. (1982), *Wilderness and the American Mind*, (Yale University Press, London).

Niva, A. (2003), *Matkailu, porojen laidunnus ja laidunten kuluminen Saariselällä: vuosien 1986 ja 2000 vertailu*, (Hämeen ammattikorkeakoulu, Evo).

Pohjois-Suomen ja Pohjois-Karjalan vanhojen metsien suojelun alue- ja kansantaloudelliset vaikutukset – valtion vanhat metsät (1995), (Jaakko Pöyry, Helsinki).

Power, T. M. (1996), *Lost landscapes and failed economies*, (Island Press, Washington D.C.).

Public Law (1964), Public law 88-577. 88th Congress. S. 4, September 3, 1964.

Saarinen, J. (1998a), 'Cultural Influence on Response to Wilderness Encounters: A Case Study from Finland', *International Journal of Wilderness*, 4: 28–32.

Saarinen, J. (1998b), 'Wilderness, Tourism Development and Sustainability: Wilderness Attitudes and Place Ethics', in A.E. Watson and G. Aplet (eds), *Personal, Societal, and Ecological Values of Wilderness: Sixth World Wilderness Congress Proceedings on Research, Management, and Allocation*, Vol. I. General Technical Report, Ogden, UT. USDA Forest Service, Rocky Mountain Research Station, 29–34.

Saarinen, J. (2003), 'The regional economics of tourism in Northern Finland: the socio-economic implications of recent tourism development and future possibilities for regional development', *Scandinavian Journal of Tourism and Hospitality*, 3: 91–113.

Saarinen, J. (2005), 'Tourism in Northern Wildernesses: Nature-Based Tourism Development in Northern Finland', in C.M Hall and S. Boyd (eds), *Nature-based Tourism in Peripheral Areas: Development or Disaster?*, (Channelview Publications, Clevedon), 36–49.

Saarinen, J. (2006), 'Traditions of Sustainability in Tourism Studies', *Annals of Tourism Research* 33(4): 1121–1140.

Saariselän yleiskaava (1993), Lapin lääninhallitus, Rovaniemi, 1–105.

Saariselän liikunta- ja virkistyspalvelut. Kehittämisohjelma 1994–2003 (1995), Metsäntutkimuslaitos ja Inarin kunta, 1–9.

Saastamoinen, O. (1982), 'Economics of the multiple-use forestry in the Saariselkä fell area', *Communicationes Instituti Forestalis Fenniea* 104: 1–102.

Seurajärvi-Kari, I., Aikio-Puoskari, U., Morottaja, M., Saressalo, L., Pentikäinen, J. and Hirvonen, V. (1995), 'The Sami People in Finland', in J. Pentikäinen and M. Hiltunen (eds) *Cultural Minorities in Finland: An Overview Towards Cultural Policy*, (Finnish National Commission for Unesco, Helsinki), 101–146.

Short, J.R. (1991), *Imagined country: society, culture and environment*, (Routledge, London).

Telfer, D.J. (2002), 'Tourism and regional development issues', in R. Sharpley and D.J. Telfer (eds) *Tourism and development: concepts and issues*, (Channel View Publications, Clevedon), 112–48.

Tourism Statistics 2004 (2005), *Transport and Tourism 2004*: 11.

Urho Kekkosen kansallispuiston hoito- ja käyttösuunnitelma (2001), *Metsähallituksen luonnonsuojelujulkaisuja* Sarja B 60:1–67.

Ympäristöministeriö (2002), 'Ohjelma luonnon virkistyskäytön ja luontomatkailun kehittämiseksi', *Suomen ympäristö* 535: 1–48.

Chapter 13

# The Economic Potential of Regional Nature Parks in Switzerland: A Case Study of the Planned Regional Nature Parks in the Canton of Bern

Dominik Siegrist, Marco Aufdereggen, Florian Lintzmeyer
and Harry Spiess

## Introduction

Beautiful and diverse landscapes have long been a feature that helped to shape Switzerland's image abroad. These landscapes are an essential part of tourism that is taken for granted. Today, nature-based landscapes in Switzerland are becoming an increasingly scarce resource.

For this reason there is a growing need among visitors for a better designation of suitable areas, including national parks, biosphere reserves, world heritage sites as well as regional nature parks and nature experience parks. Thus far, Switzerland has lacked a legal basis for these large protected areas. For this purpose, under the title 'parks of national importance', the necessary legal basis has been prepared within the framework of a revision of the federal law on the preservation of regional tradition and nature,[1] which provides for the support of the creation and funding of respective parks by federal government. The bill distinguishes three park categories: national parks, nature experience parks, and regional nature parks (see Stulz 2003, 180pp; Swiss Federal Council 2005a).

Switzerland has long had merely one classical large protected area, namely the Swiss National Park, founded in 1914 and located on 172,4 km² in the Unterengadin region. The bill specifically passed for this area defines the essence and purpose as follows:

> The national park is a reservation where nature is protected from all human interference and where fauna as well as flora are left to their natural development (National Park Law 1980).

Next to this total reservation, there are, however, two other large protected zones of national importance: moorlands and floodplains. Up to the present, Swiss legislation does not contain any basis for integrated area management of such protected areas. In recent years, due to the lack of a legal basis, mainly in the French-speaking region of

---

1   SR 451, Bundesgesetz vom 1.Juli 1966 über den Natur-und Heimatschutz (NHG).

western Switzerland, different local initiatives have arisen following the successful French concept of a Parc Naturel Régional (PNR). These park regions do not have a national legal basis and also lack legal minimum requirements. Another approach, among other things, builds on the existing national landscape protection inventory, BLN,[2] and strives for the creation of new regional nature parks with the objective of conserving and adding value to large cultural landscapes. Connected to this are also tourist and regional economic objectives that attempt to conserve the character of scenically valuable areas through appropriate use (see BUWAL 2001).

*Area Type 'Regional Nature Park'*

In the context of increasing demands for *sustainable* regional development, the 'regional nature park' is being given particular status in Switzerland (see Swiss Federal Council 2005c). The purpose for creating this new area type is to merge existing and new activities and to integrate them into a countrywide park strategy. The tools of certification and financing are essential and ought to steer the advancement of regional nature parks in a qualitative way. Hereby, different targets are addressed simultaneously, on the one hand regional economic development and on the other hand nature and landscape protection. An important aspect of participation is the federally mandated interaction of local initiative (implementation of a 'park-charter') and national assistance (certification and financing). The federal government simply wants to provide a favourable general framework for the planned parks. To be successful in the long run the projects in the regions ought to be developed from the grass-roots. Thus the chosen approach consists of a combination of a bottom-up and top-down strategies.

The federal government will only assist parks that are based on regional initiatives and that are supported by the local population. Because the federal government has declared the participation of the residents a mandatory condition for recognizing a regional nature park, this aspect gains considerable relevance in the regions. The federal government will assist the regional nature parks with financing and labels (brand-rights), whereas the rights and obligations, which are a prerequisite for the awarding of the designation 'regional nature park,' have to be arranged in an agreement between the federal government, canton and supporters. In addition, the cantons and municipalities are expected to help cover the costs of the regional nature parks. Unlike the UNESCO-biosphere reserve, no zoning takes place in a regional nature park. Nonetheless, the quality standard for nature and landscape is high and the park objectives must serve the advancement of nature and landscape (see Table 13.1). The total surface of a regional nature park can be no less than 100 km$^2$ and, in essence, has to include the undivided municipal areas (see Swiss Federal Council 2005a).

In the future the label 'regional nature park' will be awarded for the duration of ten years, to regions that stand out because of their natural, scenic and cultural features. The purpose of this is to incorporate the remarkably beautiful landscapes into the regional economic cycles and to make them useful for the local population as well as for tourism. The intention is to review and recertify the nature parks after

2   Federal Inventory of Landscape and Natural Monuments. of National Importance <http://www.umwelt-schweiz.ch/buwal/de/fachgebiete/fg_land/bln/index>.

**Table 13.1　Legal requirements for a regional nature park**

| Art. 19 Conservation and valorization of nature and landscape |
| --- |
| To conserve and enhance the quality of nature and landscape in a regional nature park particularly:<br>1. The diversity of native fauna, flora and habitats as well as the landscape and overall appearance of the locality ought to be conserved and as far as possible improved<br>2. Protection worthy habitats of local fauna and flora ought to be enhanced and linked together<br>3. New buildings and facilities ought to be built in a way that does not impair the character of the landscape and the overall appearance of the locality<br>4. Present impairments of the landscape and the overall appearance of the locality by buildings and facilities ought to be removed as much as possible. |
| **Art. 20 Economy managed in a sustainable way** |
| To promote an economy managed in a sustainable way in a regional nature park particularly:<br>1. Local and natural resources ought to be used in an environmentally sound way<br>2. Regional processing and marketing of agricultural and silvicultural park products ought to be reinforced<br>3. The use of innovative and environmentally sound technologies ought to be supported<br>4. Traditional cultivation and production methods ought to be used and developed. |

*Swiss Federal Council 2005b*

a period of ten years. During this evaluation of whether or not the objectives have been achieved, regional economic impacts will consequently be of great interest. This question will be the subject of the following case study.

*The Economic Potential of the Planned Regional Nature Parks in the Canton of Bern*

*Starting Position and Goal of the Analysis*　In this case study, we focus on the economic potential of the planned regional nature parks in the Swiss Canton of Bern (see BECO 2006). The Canton of Bern is the second most populated canton in Switzerland, with 962,000 inhabitants, and the third largest in surface area (5,959 km²). In this canton the following six regional nature park-projects have been proposed: Chasseral, Diemtigtal, Emmental, Gantrisch, Oberaargau and Thunersee-Hohgant (see Figure 13.1).

　　From our point of view, next to nature and landscape criteria, the regional economic suitabilty and hence the resulting regional economic contribution, are a significant consideration for the creation of regional nature parks. Thus, we will address the regional economic dimension of the parks and, in particular, the problem surrounding the potential of nature-based tourism and the related question of the saturation of niche markets in tourism.

　　In this context the following two questions are at the centre of our inquiry:

- What additional regional and macroeconomic effects of added value will regional nature parks trigger?
- What is the relevance of park projects from a regional point of view and which regional economic starting position and structure correlates with

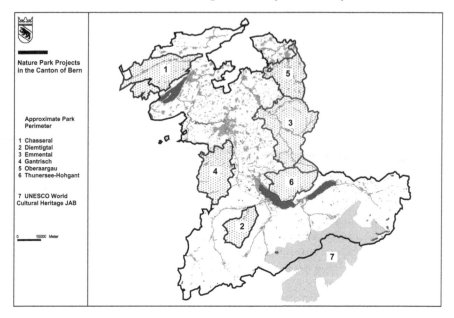

**Figure 13.1   Regional nature park projects in the Canton of Bern**
*Source: Drafted by authors*

> a particularily high suitabilty of regional nature park-projects for further economic advancement (particularly in agriculture and tourism)?

The determination of regional economic potentials and effects of regional projects and strategies is always confronted with external factors and overlapping systems that can strongly affect the results one way or another. On different regional levels there is currently a series of initiatives in progress, the implementation of which will have direct and indirect impacts on the economic progress of rural regions in Switzerland in the next few years.

For practical purposes a set of additional influencing factors could not be considered and had to be excluded from this chapter. Possible effects of future agricultural policy or other sectoral policies from the federal government on the development of nature parks had to be left out. Likewise unaccounted for were possible impacts from potential future park projects in other regions of Switzerland or similar projects that target the same demand segments in tourism.

*Description and Comparison of the Project Areas*

*Project Area Chasseral*    Chasseral is located in the southern range of the Swiss Jura between the cities of Bienne, Neuchâtel and La Chaux-de-Fonds and towers over the so-called 'Schweizer Mittelland' with its 1,607 meters above sea level. The project area Chasseral has important highmoors and swamps, which are in part listed in the Federal Inventory of Landscapes and Natural Monuments of National Importance. Furthermore, the protected amphibian spawning grounds as well as the great variety

**Table13.2    Structural data of six regional nature park projects in the Canton of Bern**

|  | Canton of Bern | Chasseral | Diemtigtal | Emmental | Gantrisch | Oberaargau | Thunersee-Hohgant |
|---|---|---|---|---|---|---|---|
| Number of Municipalities | 400 | 22 | 1 | 14 | 10 | 31 | 8 |
| Area size km² | 5,959 | 313 | 129 | 452 | 256 | 209 | 225 |
| Resident pop. 2004 | 961,647 | 24,752 | 2,049 | 37,704 | 17,028 | 30,526 | 8,728 |

of insects and birds are examples of the valuable biodiversity of this area. Chasseral constitutes the link between French and German speaking Switzerland, between the Canton of Neuchâtel and the Canton of Bern, as well as between industrialized and rural areas of the Jura Mountains. The park region is 313 km² in size and includes 17 municipalities in the Canton of Bern as well as five municipalities in the Canton of Neuchâtel (see Table 13.2).

In 1998 an initiative group composed of municipal representatives and private organizations decided to get together with the goal of developing and implementing the very first regional nature park in Switzerland. In 2001 the association *Regional Park Chasseral* was formally founded. The municipalities are financially supporting the park, but private sponsors are also already helping with specific implementation projects. With respect to tourism, Chasseral has a wide range of opportunities with numerous options for accommodation. These offers are new supplementary initiatives such as the project 'Watch Valley', which ties in with the industrial tradition of the Bernese Jura. The Chasseral-project is at an advanced stage and stands out with its large surrounding area populated with many potential tourists. One of the flaws of this regional nature park-project is, apart from the climate, the additional expenses anticipated for the administration because of the fact that the park area belongs to two cantons. At the same time, this creates an opportunity for intense cooperation between the municipalities of two different cantons and language regions.

*Project Area Diemtigtal*    Diemtigtal valley is located south-west of Lake Thun and south of the Simmental valley. The character of this multi-faceted landscape is composed of meadows, pastures, steep slopes, rockfall areas, avalanche couloirs, windfall areas and a multitude of ecological habitats. The rural landscape is characterized by a loose nuclei of populated areas as well as disperse settlements with numerous buildings in need of protection and conservation. The project area, exclusively situated within the limits of the municipality of Diemtigen, has a length of 16 km and a surface area of 130 km² which makes it the largest valley adjacent to the Simmental valley. Its surface makes Diemtigen the fifth largest municipality in the Canton of Bern.

The idea of a regional nature park Diemtigtal valley was developed by representatives of the mountain region and regional politicians. The local population considers this park project to be of great significance due to a lack of development alternatives. It goes without saying that the nature park project enjoys a high level of acceptance. The project is at an advanced stage because the park perimeter is identical with the limits of the municipality of Diemtigen and because the administrative procedures are simple. From an economic point of view the long term future of the few skiing areas is questionable. The arguments for a regional nature park Diemtigtal are the somewhat untouched potential of nature-based tourism as well as the overwhelming acceptance by the population. The weaknesses of the project are the marginal managment resources, the lack of accomodation options and the relatively small-sized area, if no cooperation with neighbouring regions takes place.

*Project Area Emmental*    The Emmental valley is situated east of the city of Bern and west of the city of Lucerne. It is adjacent to several other park projects. To the south, the municipality of Schangnau connects to the project area Thunersee-Hohgant. In the south-east the Biosphere Reserve Entlebuch is situated while in the north the project area Oberaargau is located. The natural landscape is enriched by the interplay of densly wooded rolling hills and the view of the towering Alps in the background. A portion of the project area is listed in the Federal Inventory of Landscapes and Natural Monuments of National Importance. The significant overall appearance of the localities are valuable elements of the cultural landscape. With its 452 km² and 14 diverse municipalities, Emmental is the largest project area in the Canton of Bern.

The discussion regarding a regional nature park started in 2004, and in 2005 the general public was first informed. The existing sustainable initiatives in tourism mesh well with the concept of a regional nature park. Emmental enjoys a long tradition of gastronomy, but has flaws in the quality of accomodation available. The nature park creates the opportunity to extend the famous brand 'Emmental' beyond the 'Emmental cheese' by adding value to tourism through the use of the brand. In addition, Emmental has many options for cooperation with the adjacent park regions.

*Project Area Gantrisch*    The Gantrisch region is located within the triangle of the cities of Thun, Bern and Freiburg in a hilly and mountainous area dominated by the Gantrisch range (2,188 m above sea level). Moorlands and swamps are the most important natural features, some of which belong to the Federal Inventory of Landscapes and Natural Monuments of National Importance. The project area Gantrisch is characterized by a rich cultural landscape with disperse settlements, cultivated land and small forests with wild canyons. The total surface area is 316 km² and includes ten municipalities.

The project was launched by proactive organizers from the region in 2004 starting from the regional development project 'Wald Landschaft Gantrisch'. The main opportunities for a successful nature park stem from the close proximity of the recreational area to the city of Bern as well as the potential concentration of regional resources. The population has a relatively large interest in the nature park project and furthermore, the municipalities have already helped fund various implementation projects. However, from a tourism point of view Gantrisch lacks a real center and there are few accomodation options of higher quality.

While opportunities to experience nature and the successful regional product label fit into the concept of a regional nature park, the small tourism infrastructure as well as potential conflicts with other forms of use in the area adjacent to the agglomeration could be considered potential weaknesses.

*Project Area Oberaargau*    The region Oberaargau located within the triangle of the cities of Lucerne, Bern and Aarau is the furthest northeastern part of the Canton of Bern. The landscape on the foothills of the Alps is heavily sculpted by former glaciers; moraines, areas of gravel, boulders, lake and moor areas are typical in this part of the canton. A special highlight is the cultural landscape 'Wässermatten', where with federal financial support the tradition-rural cultivation is kept alive by means of a special watering system. The project perimeter comprises on a territory of 210 km² a total of 31 municipalities and part of the project area is listed in the Federal Inventory of Landscapes and Natural Monuments of National Importance.

A few years ago, due to the initiative of a private nature protection organization, the park project was adopted into the development concept, even though the project region is not exceptionally marginal. In fact, Oberaargau is a popular local recreational area with a large belt of potential tourists in the surrounding areas and an exquisite gastronomy. As a result of its low differences in elevation, the area is especially suited for hiking trips for all age groups.

*Project Area Thunersee-Hohgant*    The project area Thunersee-Hohgant is situated between the cities of Thun, Spiez and Brienz above the right hand side of the shore of Lake Thun. Among the most essential natural values of this very scenic area are the moorlands which hold national importance. Parts of the project area are listed in the Federal Inventory of Landscapes and Natural Monuments of National Importance. From the peak of Hohgant (2197 m above sea level) down to Lake Thun at approximately 560 m above sea level, the project area covers a surface of 225 km². The area includes eight municipalities with some large economic disparities.

In 2000 the discussion about a regional nature park project began under the leadership of the tourism community of Beatenberg. In 2001 the initiative was reinforced with the founding of an association and the launching of the project 'Höhenweg Thunersee'. The nature park project stands out because of the multitude of implementation projects that are already in effect. The project also offers the opportunity to cooperate with the Emmental project as well as with the Entlebuch project. The project area is located in the surroundings of the Lake Thun destination and has a well equipped tourism infrastructure with a wide range of opportunities. In addition to the regional nature park there are also other development alternatives.

## Evaluation of Regional Economic Potential

### Preliminary Remark

Below, based on a regional economic model calculation, an evaluation is conducted as to the potential regional effects of added value in nature park projects as well as to the demand potential of regional nature park tourism.

In this model (see Figure 13.2) it is assumed that for a regional nature park specific regional inflow and outflow of funds can be accounted for. For now all other cash flow is excluded, that is processes in the context of the regional nature park are considered to be isolated from the total system.

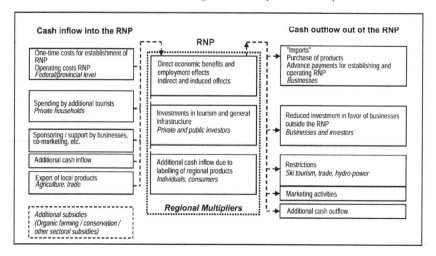

**Figure 13.2   Cash in- and outflows of a regional nature park**
*Source: Drafted by authors after Getzner, Jost and Jungmeier 2001, 13*

The variables quoted in the figure are accounted for as follows:
Funds inflow:

- One-time setup costs and operating expenses of regional nature parks: Future anticipated contributions by the cantonal and the federal government. These contributions were differentiated according to the size of each project area. They will amount to 75 per cent of the parks' entire operating expenses
- Number and expenditures of additional visitors
- Endorsement/support by companies, co-marketing: assumption of future support by the private sector and endorsements.

Due to a lack of data, other inflowing funds, export of local products or additional subsidies (agriculture, nature protection) are not accounted for in the model calculation.

With regard to number and expenditures of additional visitors[3] and endorsements,[4] there are three possible scenarios envisioned that employ different assumptions with regard to the actual amount or ratio of nature based tourists, their future increase, their additional expenses and the anticipated amount of funding from sponsorship (see SECO 2003) (see Figure 13.3).

---

3   The average expenditures per day for overnight accommodation and single day visitors in the Canton of Bern are based on a study researching the added value in tourism Müller, Rütter and Guhl (1995, 30). With the cost adjusted for inflation 1996-2005 (+107%) the expenditures amount to 166 SFr for the current hotel guest in Berner Oberland, 75 SFr for the overnight guest in accommodation other than hotel and 56 SFr for the guest on a day trip; respectively, in the region Mittelland 222/44/48 SFr and in the Jura region 130/43/33 SFr and for the canton as a whole 177/66/50 SFr.

4   Based on the actual amount and the anticipated endorsements contributing to the operating expenses of nature parks in Austria; oral information by the director of the Nature Park Association Austria.

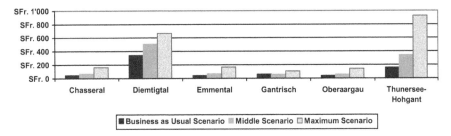

**Figure 13.3  Assessment of per-capita added value due to a regional nature park**

*Source: Drafted by authors*

*Regional Multipliers*

Every expenditure by a guest in the project areas leads through several added value chains to an increase in sales volume for the suppliers of tourism businesses (indirect effects) as well as to an increase of regional incomes and therefore in turn to a further raise of the sales volume triggered by increased consumption (induced effects). These rather complex procedures are presented in regional added value analyses by means of multpliers; one can assume a multiplier of 1.4 to 1.7 for the aggregated added value on the basis of revenues generated in the tourism region.[5] For the estimation at hand, a multiplier across the board for all regions of 1.5 was chosen, because a differentiation would have required the support of much more extensive field research. Depending on regional pre-conditions and efforts to tie the value chains closer to the region, in sum, this factor could be even higher.

*Outflow of Funds*

Those inputs needed for tourism from outside the region (know-how, preliminary products, workforce, etc.) will be deducted as outflow of funds. To simplify the calculation of the amount of these inputs, we use the 'production account of Switzerland' from the Federal Statistical Office, which asseses for the year 2003 the amount of inputs as part of the gross production value on an average of 45 per cent (BFS 2005). This amount will be, with the calculation at hand, deducted from the added value induced by the nature park.

*Scenarios*

Three scenarios of the future development of nature park tourism in the Canton of Bern were selected, because the range of unknown future framework conditions and demand trends in tourism ought to be left open (see Table 13.3).[6]

---

5   See. Buser, Buchli and Giuliani (2003, 5) for the Val Müstair; Simeon, Zgraggen and Koch (2004, 45) for the nature park Ela; Guhl in Buehler and Minsch (2004, 23 et seq.) for the Canton of Grisons; Müller, Rütter and Guhl et al (1995, 69) for the Canton of Bern.

6   Cf. Spiess, von Allmen and Weiss-Sampietro 2005.

**Table 13.3    Development scenarios on tourism in regional nature parks in the Canton of Bern**

| |
|---|
| Scenario 'Business as Usual' |
| In the 'Business as Usual' scenario the following assumptions were made: |
| The market share of the guests with the preference 'regional nature park' is at 7 percent[7] of all guests and visitors that spend their leisure time in the Canton of Bern. |
| Until 2015 the number of guests with the preference 'regional nature park' will increase by 10 percent of today's share to a market share of 7.7 percent. |
| In 2015 the willingness to spend more money will not have increased from the current level. |
| The level of sponsorship will amount to 5 percent of the operating expenses of the regional nature park. |
| Scenario 'Medium' |
| In the 'Medium' scenario the following assumptions were made: |
| The market share of the guests with the preference 'regional nature park' is at 15 percent of all guests and visitors that spend their leisure time in the Canton of Bern. |
| Until 2015 the number of guests with the preference 'regional nature park' will increase by 20 percent of today's share to a market share of 18 percent. |
| In 2015 the willingness to spend more money will have increased by 10 percent from the current level. |
| The level of sponsorship will amount to 10 percent of the operating expenses of the regional nature park. |
| Scenario 'Maximum' |
| In the scenario 'Maximum' the following assumptions were made: |
| The market share of the guests with the preference 'regional nature park' is at 30 percent of all guests and visitors that spend their leisure time in the Canton of Bern.[8] |
| Until 2015 the number of guests with the preference 'regional nature park' will increase by 40 percent of today's share to a market share of 42 percent. |
| In 2015 the willingness to spend more money will have increased by 10 percent from the current level. |
| The level of sponsorship will amount to 10 percent of the operating expenses of the regional nature park. |

## Results

All project areas are expected to have additional effects of added value with regional nature parks. These effects can vary depending on the project area and the scenario (see Table 13.4).

*Chasseral*

With a small project area, there will be per capita, small additional effects of added value; in an absolute way, there will be small to medium effects by the regional

7    The World Tourism Organization (WTO) estimates the amount of expenditures in ecotourism as compared to the market as whole with 7 percent (Lindberg, Furze and Staff et al 1997). For the calculation at hand applies the assumption: 7 percent expenditure = 7 percent market share.

8    Cf. SECO 2002, 36 and http://www.invent-tourismus.de.

nature park, even though there is a larger tourism infrastructure. Because there are no real opportunities for expansion, the additional potentials of added value lie primarily in regional cooperation.

*Diemtigtal*

Depending on the scenario, the additional effects of added value by the regional nature park are expected to be medium to large per capita and small in an absolute sense, even though it is a small project area with a small tourist accommodation infrastructure. It is a prerequisite to have sustainable park management and tourism marketing, which raises the question of an expansion of the project area as well as cooperation with the neighbouring regions.

*Emmental*

Despite its large project area and the existing tourist offerings, the additional effects of added value by the regional nature park are expected to be small per capita and small to large in an absolute sense, because of the large number of inhabitants in Emmental. Additonal potential for this project lie in the connection of the projects Emmental and Thunersee-Hohgant as well as in the cooperation with the adjacent UNESCO-biosphere Entlebuch in the Canton of Lucerne. The Emmental community of Schangnau currently participates in both projects: Emmental and Thunersee-Hohgant.

*Gantrisch*

The additional effects of added value by the regional nature park are expected to be small per capita and in an absolute sense, because, along with a large number of inhabitants is only a small number of tourist offerings. In the medium-term though, there is additional potential in connecting the projects Gantrisch and Diemtigtal (with the inclusion of Simmental).

*Oberaargau*

The additional effects of added value by the regional nature park are expected to be small per capita and in an absolute sense, because there exists only a limited number of tourist offerings next to a large number of inhabitants.

*Thunersee-Hohgant*

Depending on the scenario, the additional effects of added value by the regional nature park are expected to be small to high per capita and in an absolute sense, because there is a broad tourist infrastructure. Additonal potential remains in the connection of the projects Thunersee-Hohgant and Emmental as well as in the cooperation with the adjacent UNESCO Biosphere Entlebuch in the Canton of Lucerne.

**Table 13.4    Comparative assessment of potential additional regional added value due to regional nature parks in the Canton of Bern in 2015**

| Scenario | Chasseral | Diemtigtal | Emmental | Gantrisch | Oberaargau | Thunersee-Hohgant |
|---|---|---|---|---|---|---|
| Scenario A absolute | • | • | • | • | • | • |
| Scenario A per capita | • | •• | • | • | • | • |
| Scenario B absolute | •• | • | ••• | • | • | ••• |
| Scenario B per capita | • | ••• | • | • | • | ••• |
| | Minor potential for additional added value | | | | | • |
| Legend | Medium potential for additional added value | | | | | •• |
| | Considerable potential for additional added value | | | | | ••• |
| Scenario A: 'Business as usual'/Scenario B: 'Maximum' | | | | | | |

*Regional Economic Potential*

By dividing the additionally expected effects of added value by the regional nature park through the number of inhabitants in the project area, one receives the per capita effect of the regional nature park. Due to the small number of inhabitants, the project area Diemtigtal scores the highest, followed by Thunersee-Hohgant and with some distance Emmental and the remaining project areas (see Figure 13.3 and Table 13.4).

The regional nature park may unfold a medium to large regional economic effect in the project areas depending on the starting position. This will strengthen the position of the regional suppliers on very specific niche markets. The increase in competitiveness can lead, under favorable conditions, to a decreasing dependency of nature park regions on public funding.

It becomes evident what significant regional economic importance a regional nature park can have for some areas, in this case, for Diemtigtal. In order to realize the regional economic potential, the issue of the meager management resources needs to be resolved. If one assumes, however, that an effect recognizable by the population will lead to an increase in the level of acceptance and consequently the identification with the regional nature park, then one would have to conclude that the per capita-effect could be an argument for a limitation in the size of each park region.

*Overall Economic Potential*

The situation presents itself differently when assessing the absolute potential added value in the region of the park projects. For the overall economic potential, only the expected effects of added value from the larger project areas with a respective economic volume are relevant (see Figure 13.4).

Due to the large number of accommodation in the project area Thunersee-Hohgant and due to a high level of assistance and the significant number of daytime tourists in Emmental, these two areas are in the lead (see Figure 13.4). Chasseral and Oberaargau follow, while Gantrisch and Diemtigtal are in last place with regard to the absolute economic potential added value.

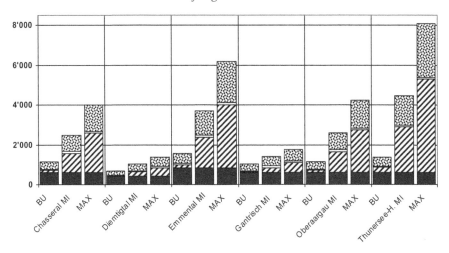

| Project Area | Potential Effect | Project Area | Potential Effect |
|---|---|---|---|
| Chasseral BU | 1,147,000 SFr. | Gantrisch BU | 1,029,000 SFr. |
| Chasseral MI | 2,489,000 SFr. | Gantrisch MI | 1,403,000 SFr. |
| Chasseral MAX | 3,990,000 SFr. | Gantrisch MAX | 1,768,000 SFr. |
| Diemtigtal BU | 698,000 SFr. | Oberaargau BU | 1,159,000 SFr. |
| Diemtigtal MI | 1,032,000 SFr. | Oberaargau MI | 2,607,000 SFr. |
| Diemtigtal MAX | 1,367,000 SFr. | Oberaargau MAX | 4,229,000 SFr. |
| Emmental BU | 1,574,000 SFr. | Thunersee-Hohgant BU | 1,369,000 SFr. |
| Emmental MI | 3,713,000 SFr. | Thunersee-Hohgant MI | 4,464,000 SFr. |
| Emmental MAX | 6,167,000 SFr. | Thunersee-Hohgant MAX | 8,076,000 SFr. |

*BU = Business as Usual-Scenario*     *MI = Middle Scenario*
*MAX = Maximum Scenario*

**Figure 13.4   Assessment of potential regional added value due to regional nature parks in the Canton of Bern**

*Source: Drafted by authors*

Generally speaking, the overall economic potential of all nature park-projects in the Canton of Bern ought to be classified as rather low. Indeed the regional nature parks will probably induce additional effects of added value. However, in the context of the overall economy of this large Swiss canton, these effects will be covered up by other economic processes and therefore, they can hardly be statistically isolated. Thus, the regional nature parks are more likely to play the role of a catalyst for tourism and rural regional development in the Canton of Bern, than the engine of the overall economy.

*Regional Development Alternatives*

The quantified estimate of overall and regional economic potential can present an important basis for the integrated evaluation of regional nature park-projects.

This potential, however, must be complemented by an examination of the regional development perspectives without a regional nature park. This may vary greatly depending on the socio-economic and geographic starting position.

Only three out of the six project regions show sound development alternatives without a regional nature park: Thunersee-Hohgant with its tourist potential, Oberaargau as a commuter region close to the strong economic region Schweizer Mittelland, as well as Chasseral with its relatively broad industrial foundation.

Emmental, with its large number of purely agrarian communities, has average development alternatives without a regional nature park. Gantrisch and Diemtigtal possess few alternatives due to the high employment rate in the primary sector and due to the peripheral location.

## Potential and Opportunities in Tourism

For most project executing organizations an important motivation for their commitment in nature park development is to open up new tourist (niche) markets. Hence, it is essential to answer the question of the potential and opportunities of nature park tourism. This becomes especially important if one keeps in mind the increasing competition in tourism among the alpine countries, which face the problem of raising saturation of the market and stagnation in various segments.

In the future the expected demand for nature park-tourism will depend on two factors: first, the general tourism trends that can hardly be influenced by policy measures and second, the quality of tourism and the quantitative amplitude of nature park-offerings. Important factors for success are a positioning that is based on real regional strengths as well as an intelligent demand differentiation between regional nature parks and comparable offers. Furthermore, the risk of a saturation of the market by an excessive amount of regional nature parks can be decreased by using a substantiated regional and national nature park-strategy.

According to the literature, the market share of nature-based tourists is estimated between 7 and 30 per cent. In addition, survey-based forecasts estimate the growth rate in nature-based tourism between 10 and 40 per cent within ten years. In tourism circles nature-based tourism is seen as having growth potential. This assumption can be justified by the fact that the offers in the nature parks are not ideally developed yet and the efficiency of marketing needs to be improved (see SECO 2002, Leuthold 2001).

The entire potential of nature-based tourism in the Canton of Bern (scenario 'medium') is at roughly 1.3 million guest-days. If one compares this number with the additional guest potential of the Bernese nature park-projects, it becomes clear that it will not approach the demand potential of the future. It must be noted that, as a limiting factor, large tourist resorts will increasingly target the segment of nature-based tourism and hence, compete with regional nature parks on the tourism market.

## Conclusions

Starting from the discussion about the new area category regional nature park in Switzerland, we have addressed the question of regional economic potential of such areas by means of a case study. In the process of this it was determined that additional effects of regional added value can be expected within the examined nature park-projects. These effects can vary greatly depending on the project area and the expected basic conditions. The actual exhaustion of these potentials relies greatly on the successful integration of the regional nature parks into the marketing of tourist destinations.

With regard to the future development of new parks, the question arises as to where the different objectives between protection and development can meet and where they are presumably going to be exclusive. Most regional players agree that there is potential for synergy between the promotion of the regional economy and nature protection (promotion of ecological agriculture, creation of nature-based tourist offerings, improvement of regional cooperation, etc.). Whether these synergies will actually be utilized hinges on whether or not the factors of success and failure are taken into account in the planning and implementing strategies of the different regional levels (see Siegrist 2004). Thus an essential issue concerns the cooperation between the different regional players, especially between nature park management, municipalities, and key players in tourism, agriculture and nature protection. In this environment the nature park administration can play the role of a facilitator and use its medial position in the search for sustainable solutions within the nature park region.

Diverging interests are noticeable among the various stakeholders with regard to the regional economic level. There, the problem is that the amplitude of the contribution from new nature parks towards the regional economy is oftentimes overrated, particularly with regard to the direct and indirect effects of added value. Where the announced economic benefit by the new parks does not happen, the desire for new investments, infrastructures and activities that are not necessarily compatible with nature and landscape could become stronger. This could then lead to a situation of new and unwanted conflict between nature protection and economic development.

Whether the expected synergetic effects will actually happen within the framework of the new regional nature parks will hinge greatly on whether or not the stakeholders and promoters are in a position to recognize the potential and also whether the prospective conflicts are identified and integrated early on into their own strategies. This is especially true for tourism that expects new momentum from nature parks in light of the Swiss-saturated tourism market. Consequently, the umbrella organization 'Swiss Tourism' increased the promotion of nature-based tourism products in recent years. Not just marketing organizations in tourism, but also a number of nature protection organizations together with tourist professionals have started to shape and market exemplary nature-based tourist offerings in the past years. Such strategies focus on the predicted leisure trend, a continuously growing demand in the nature-based segment of the market which has emerged slowly for some time now.

Not all regional nature parks can benefit to the same degree from the expected positive trend in demand in nature-based tourism. Because many project areas are within day-trip-distance of the populous centres of Switzerland, and therefore register a high share of guest spending for a single day, the result is low regional effects of added value. In some regions there are already diversified tourist offerings with strong revenues, in other regions this is less frequently the case. The room and board infrastructure varies greatly quantitatively and qualitatively, and therefore a possible designation as regional nature park will not have the effect of a revenue engine in all regions. The general development context of each respective project region is, however, also relevant. While in some areas the regional nature park forms a basic element of the regional development strategy, in other areas there are further development alternatives. In this sense, the designation of new regional nature parks is not just a regional economic challenge, but also a regional policy challenge.

From an overall economic perspective, the potential of nature park-projects is rather low. Seen from a distance, the regional nature parks in Switzerland will not play the role of an economic engine, but rather the one of a catalyst. The situation is fundamentally different for the peripheral regions. Depending on the starting point, the peripheral regional nature parks can develop significant regional economic effects. In the end, this can result in decisive contributions towards the development of weak economic regions.

## References

AGR (2006), *NHG Teilrevision: Pärke im Kanton Bern*. Abteilung Kantonsplanung (Bern).

BAK Basel Economics (2003), *Der Tagestourismus in der Schweiz* (Basel).

Beer, F. (2005), *Der mögliche Beitrag von Regionalen Naturparks zur touristischen Wertschöpfung in der Schweiz* (Samedan).

Broggi, M.F., Staub, R. and Ruffini, F.F. (1999), *Grossflächige Schutzgebiete im Alpenraum: Daten. Fakten. Hintergründe* (Bozen: Blackwell Publishing).

Bühler, D. and Minsch, R. (2004), *Der Tourismus im Kanton Graubünden: Wertschöpfungsstudie* (Chur).

BECO (2006). Ökonomische Analyse von regionalen Naturpark-Projekten im Kanton Bern. Dominik Siegrist, Marco Aufdereggen, Florian Lintzmeyer, Forschungsstelle für Freizeit, Tourismus und Landschaft der Hochschule für Technik Rapperswil, Harry Spiess, Institut für Nachhaltige Entwicklung der Zürcher Hochschule Winterthur im Auftrag der Berner Wirtschaft beco (Bern).

BFS (2005), 'Produktionskonto der Schweiz', *BFS aktuell, Volkswirtschaft 4* (Neuenburg).

Buser, B. (2005), *Regionale Wirtschaftskreisläufe und regionale Wachstumspolitik* (Aachen: Shaker Publishing).

Buser, B., Buchli, S., Giuliani, G. and Rieder, P. (2003), 'MovingAlps – Blick in die Wirtschaft einer Bergregion', in: *Montagna* 2003:3, 1–8.

Bundesrat (2005), *Botschaft zur Teilrevision des Natur- und Heimatschutzgesetzes* 05.027, dated 02/23/2005 (Bern).

BUWAL (2001), *BLN-Land Schweiz: Synergien zwischen Landschaftsschutz und Tourismus*, Werkstattbericht (Bern).

Carabias-Hütter, V. and Renner, E. (2004), *Indikatoren: Nachhaltige Regionalentwicklung verstehen, messen, bewerten und steuern*, NFP 48, Landschaften und Lebensräume der Alpen, Projekt FUNalpin 5 (St. Gallen).

Carabias-Hütter, V., Kümin, D., Siegrist, D. and Wasem, K. (2005), *Zertifizierung: Konzept für einen indikatorenbasierten Zertifizierungsprozess von Bergregionen*,

NFP 48, Landschaften und Lebensräume der Alpen, Projekt FUNalpin 8 (St. Gallen).

Getzner, M., Jost, S. and Jungmeier, M. (2001), *Regionalwirtschaftliche Auswirkungen von Natura 2000-Schutzgebieten in Österreich* (Klagenfurt).

Hammer, T. (2005), 'Tatort "Schutzgebiet": Handeln für eine nachhaltige Regionalentwicklung' in: *Alpine Raumordnung* 26, 37–47.

Hammer, T. (Ed.) (2003), *Grossschutzgebiete: Instrumente nachhaltiger Regionalentwicklung* (München: OEKOM).

Job, H., Harrer, B. and Metzler, D. (2005), 'Ökonomische Effekte von Großschutzgebieten', *BfN-Skripten* 135 (Bonn).

Job, H., Metzler, D. and Vogt, L. (2003), 'Inwertsetzung alpiner Nationalparks: Eine regionalwirtschaftliche Analyse des Tourismus im Alpenpark Berchtesgaden' *Münchner Studien zur Wirtschafts- und Sozialgeographie* 43 (Regensburg).

Krietemeyer, H. (1983), 'Der Erklärungsgehalt der Exportbasistheorie', *Schriften des Zentrums für regionale Entwicklungsforschung der Justus-Liebig-Universität Gießen* 25 (Hamburg).

Küpfer, I. (2000), 'Wieviel verdient die Region am Nationalparktourismus?' in *Cratschla* 2/2000, 10–5.

Lindberg, K., Furze, B. and Staff, M. (1997), *Ecotourism in the Asia-Pacific Region: Issues and Outlook* (Burlington).

Mose, I. (2005), 'Integrierte ländliche Entwicklung in Europa: neue Entwicklungsperspektiven für die "extreme Peripherie"?' in: *Alpine Raumordnung* 26, 19–30.

Mose, I. and Weixlbaumer, N. (Hrsg.) (2002), *Naturschutz: Grossschutzgebiete und Regionalentwicklung* (Sankt Augustin: Akademia-Verlag).

Regierung des Kantons Bern (2004), *Weiterentwicklung der Wachstumsstrategie* (Bericht des Regierungsrates an den Grossen Rat, dated 02/11/2004) (Bern).

Regierung des Kantons Bern (2005), *Gesetz zur Förderung von Gemeindezusammenschlüssen im Kanton Bern* (effective as of 6/1/2005) (Bern).

Regierung des Kantons Bern (2005), *Strategie zur differenzierten Stärkung des ländlichen Raums im Kanton Bern* (Bericht des Regierungsrats an den Grossen Rat, dated 10/19/2005) (Bern).

Regierungsrat des Kantons Bern (2005), *Strategie für Agglomerationen und regionale Zusammenarbeit im Kanton Bern* (Umsetzvorlage zur Vernehmlassung, dated 11/05/2005) (Bern).

Rieder, P. and Buser, B. (2005), 'Zur Wirtschaftsstruktur von Graubünden' in: *Bündner Monatsblatt*, 2/2005, 129–51.

Rüegg, E. (2004), *NFP 48 Landschaften und Lebensräume der Alpen: Wie macht man Alpine Landschaften zum Entwicklungsfaktor?* (Brugg).

Rütter, H., Guhl, D. and Müller, H. (1996), *Wertschöpfung Tourismus: Ein Leitfaden zur Berechnung der touristischen Gesamtnachfrage* (Bern).

Rütter, H., Müller, H., Guhl, D. and Stettler, J. (1995), *Tourismus im Kanton Bern* (Bern).

Schätzl, L. (2003), *Wirtschaftsgeographie 1: Theorie* (Paderborn: UTB).

Schweizerischer Bundesrat (2005a), *Botschaft zur Teilrevision des Natur- und Heimatschutzgesetzes 05.027*, dated 02/23/2005 (Bern).

Schweizerischer Bundesrat (2005b), *Verordnung über die Pärke von nationaler Bedeutung* (Pärkeverordnung, PäV), draft version dated 10/18/2005 (Bern).

Schweizerischer Bundesrat (2005c), *Botschaft über die Neue Regionalpolitik 05.080* (NRP), dated 11/16/2005 (Bern).

SECO (2002), *Naturnaher Tourismus in der Schweiz: Angebot, Nachfrage und Erfolgsfaktoren*. Dominik Siegrist, Silvia Stuppäck, Forschungsstelle für Freizeit, Tourismus und Landschaft der Hochschule Rapperswil, Hans-Joachim Mosler, Robert Tobias, Abteilung Sozialpsychologie II der Universität Zürich im Auftrag des Staatssekretariats für Wirtschaft Seco (Bern).

Siegrist, D. (2005), 'Erfolgsfaktoren für ein nachhaltiges Tourismusmanagement in Naturparken' in: *Alpine Raumordnung* 26, 31–36.

Siegrist, D. (2004), 'Sustainable tourism and large protected areas. Analysis models and success criteria of a sustainable tourism management using the example of the Alps' in: Sievänen, T., Erkkonen, J., Jokimäki, J., Saarinen, J., Tuulentie, S. and Virtanen, E. (eds.), *Policies, Methods and Tools for Visitor Management*, Proceedings of the Second International Conference on Monitoring and Management of Visitor Flows in Recreational and Protected Areas, June 16–20, 2004, Rovaniemi, Finland, Workingpapers of the Finnish Forest Research Institute, 319–25.

Siegrist, D. (2003), 'Optimisation and monitoring of the recreational value of natural forests: The example of the Alps' in: Hamor, F.D.; Commarmot, B. (eds): *Natural Forests in the Temperate Zone of Europe – Values and Utilisation*, Conference Proceedings, International Conference in Mukachevo, October 13–17, 2003 Swiss Federal Research Institute WSL (Birmensdorf), 311–17.

Siegrist, D. (2002), 'Naturnahe Kulturlandschaften als Ausgangsbasis für Regionalparke in der Schweiz' in: Mose, I. and Weixlbaumer, N. (Eds): *Naturschutz: Grossschutzgebiete und Regionalentwicklung* (St. Augustin: Akademia-Verlag), 155–92.

Siegrist, D. and Lintzmeyer, F. (2006). Wertschöpfungspotenzial und gesamtwirtschaftliche Bedeutung von Pärken. *Schweizerische Gesellschaft für Agrarwirtschaft und Agrarsoziologie*. 1/2006, 127–41.

Spiess, H., von Allmen, M. and Weiss-Sampietro, T. (2005), *Szenarien*, NFP 48, Landschaften und Lebensräume der Alpen, Projekt FUNalpin 6 (St. Gallen).

Weixlbaumer, N. (2005), 'Zum Mensch-Natur-Verhältnis: Naturparke als Innovationsfaktoren für Ländliche Räume' in: *Alpine Raumordnung* 26, 7–18.

Weixlbaumer, N. (2002), 'Die Chance liegt in der Umsetzung des Anspruchs; Gebietsschutz und Regionalentwicklung? Das Beispiel des Regionalparks Grands Causses (Massif Central)' in: Mose, I. and Weixlbaumer, N., *Naturschutz: Grossschutzgebiete und Regionalentwicklung* (St. Augustin: Akademia-Verlag), 193–215.

# PART III
# Synthesis

Chapter 14

# Protected Areas and Regional Development in Europe: Towards a New Model for the 21st Century

Thomas Hammer, Ingo Mose, Dominik Siegrist and Norbert Weixlbaumer

Area protection policy is currently experiencing a remarkable growth in significance in almost all European countries. This is manifest not only in the creation of new protected areas and the expansion of existing ones, but also becomes apparent in the ongoing paradigmatic debate in research and practice on the significance of protected areas and their potential role as instruments for regional development. This is particularly true in the case of large protected areas which are of major interest on the basis of their spatial extent alone. Furthermore, the greater majority of large protected areas are in peripheral rural areas which are frequently characterized by a lack of suitable development potential. Such large protected areas are increasingly assigned a role in the development of these regions, and are often created with this express intention.

The recent development of large protected areas in Europe is difficult to summarize briefly. This is due to several factors: firstly, the number of large protected areas is now so great that an up-to-date overview of the overall situation that goes beyond mere statistical data is almost impossible. Secondly, the large protected areas have undergone a remarkable differentiation of types and categories which makes it necessary to differentiate carefully between the various approaches to the development of protected areas. As well as the administrations of the protected areas, new actors are increasingly involved in their development and use is made of diverse subsidy programmes which further contribute to the development of marked 'obscurity' (*Unübersichtlichkeit*) in protected area policy, to use a concept introduced by Habermas (1998).

The case studies of eleven large protected areas presented here were selected with the specific aim of enabling a comparative, cross-section examination of different types of protected area at an European level. Although the individual case studies in themselves provide interesting insights into planning and development practice for protected areas, a generalising synthesis is only possible with the comparative approach. This method is oriented in principle towards the 'multiple case studies' methodological approach described by Yin (2003), although it must be conceded that the case studies used here use standardized methods of data collection and evaluation only to a limited extent. This was, however, not possible in the framework of the publication concept chosen for the preparation of this volume of essays.

Seven questions of overriding importance were defined for the evaluation of the case studies, the answers to which should help to pin down generalizable conclusions. These aim in particular to make specific statements on the scope and quality of the integration of area protection and regional development.

In the following, the results of the cross-section analysis are briefly sketched under the headings of the questions posed.

## Question 1: In What Kind of Environment (Geographic, Economic, Social, etc.) are Large Protected Areas Regarded as being Tools for Regional (Rural) Development?

The target areas in which protected areas can be seen as instruments of regional development are closely linked to the primary purpose of the respective categories of the protected area. Thus the national parks are by definition a tool for environmental protection, while the nature parks and to an extent the biosphere reserves are seen more as instruments for sustainable regional development in rural areas. However, there are marked differences from country to country, which can often be explained by the course of development of protected areas specific to the individual countries. This is particularly true for Eastern Europe as well as for some southern European countries (see chapters 4 and 9).

The postulated paradigm shift in the debate on protected areas is quite clearly confirmed by most of the case studies. The new approach links environmental protection with the encouragement of sustainable regional development. This discussion seems to be furthest advanced in the states of Western and Northern Europe where all categories of large protected areas are perceived as instruments for regional development (and actually fulfil this function) (see chapters 5, 6, 10 and 11). Emphasis is thereby placed on the preservation of traditional forms of land-use and the maintenance of associated cultural traditions, as well as the encouragement of close-to-nature tourism and the marketing of regional products. The regional policy background to the changed paradigm is shaped by the search for alternative forms of development for rural regions which are particularly affected by global structural change. As an instrument to accompany socio-economic transition in peripheral regions, large protected areas can take on an important function here (see chapters 12 and 13).

However, stagnation of the regional economy is not always the reason for the foundation of new parks. Dynamic development and a rapid degregation of the landscape can also be the starting point for new large protected areas. Often, individuals' resistance and social movements against the threatened destruction of natural heritage lead to the planning and creation of large protected areas. Thus, in mountain areas hydroelectric projects or plans for skiing facilities sometimes shape the history of park projects, in other areas planned holiday resorts or military use prompt their development (see chapters 8 and 7).

**Table 14.1  Areas in which large protected areas are seen as an instrument for regional development**

| Area | Examples |
|---|---|
| Regional development in its original meaning | Regional marketing and development of regional economic cycles, regional co-ordination and co-operation, participation in regional development programmes |
| Tourism | 'Close-to-nature' tourism and ecotourism, implementation of regional tourism strategies, cushioning the impact of regional structural transformation, Environmental Education |
| Traditional land-use | Maintenance and modernization of agriculture, boosting of sustainable forestry |
| Culture | Preservation of traditional customs and cultural traditions, linking with modern culture |
| Nature protection and landscape planning | Encouragement of biodiversity, preservation of forest habitats, cultural landscape planning |

## Question 2: How is this Objective – Large Protected Areas as Tools for Regional (Rural) Development – Reflected in Relevant Planning or Development Concepts?

With their claim to be model regions for sustainable development, large protected areas need suitable legal foundations, concepts, instrument and financing at different spatial and sectoral levels. Most large protected areas are based on a network of concepts for regional development and planning which make it possible for the parks to function in a regional context in the first place. The intelligent linking of basic foundations and different concepts can make it possible for protected areas to make an effective contribution to sustainable regional development (see chapters 7 and 11).

The presence of a legal basis is a central precondition for the functioning of large protected areas, and thus for their contribution in the context of regional development. This basis consists firstly of the national laws for environmental protection, national parks and protected areas, on which the relevant international agreements, guidelines and conventions are based. The parallel structures of regional development and planning laws also have considerable significance; these are often regionally specific, but also sometimes operate at EU level (see chapters 7 and 9).

Concepts oriented towards regional development support and encourage the socioeconomic and cultural development of regions in the context of the respective large protected areas. Regional development plans and associated financial instruments as well as sectoral concepts such as regional tourism programmes and regional marketing concepts are of major significance. Regional development concepts are often based on incentive-oriented strategies to motivate the actors, and especially the economic stakeholders, to engage in regionally oriented behaviour (see chapters 10 and 12).

These planning concepts are characterized by the a priori, forward-looking co-ordination of spatially effective measures (see Mönnecke and Wasem 2005). By establishing aims and indicating possible solutions, guidelines for different types of use for an area are established. These guidelines establish ways in which the quality of

**Table 14.2    Structures and concepts of large protected areas that are relevant to regional development**

| Structures and concepts | Examples |
| --- | --- |
| Legal structures | International agreements and conventions |
| | Planning laws (EU, national, regional) |
| | Regional development laws (EU, national, regional) |
| | Environmental protection laws and guidelines (EU, national, regional) |
| | National park laws |
| Regional development concepts | EU programmes (e.g. LEADER, INTERREG) |
| | National and regional development plans |
| | Regional tourism concepts |
| | Regional marketing concepts |
| Planning concepts | EU programmes (e.g. LIFE, NATURA 2000) |
| | Regional development plans and programmes |
| | National park plans |
| | Protected area management plans |
| | Network strategies |
| | Concepts for environmental education and visitor management |
| | National and regional environmental protection strategies |
| | Zoning concepts for large protected areas |

an area can be assured. The relevant planning concepts include regional development plans, national park plans and management concepts for the protected areas. The international networking of large protected areas is of growing significance. While planning concepts are often traditionally based on standard strategies controlled by the authorities, the newer generation of concepts is increasingly shifting towards persuasive and participative strategies (see chapters 3 and 6).

A trend common to many regional development and regional planning concepts is their bottom-up orientation. The growing importance of participation by the local population and regional stakeholders in regional and protected area development is attributable not least to the increased involvement of protected areas and regions in EU regional development programmes, where participative approaches are often very important. Bottom-up approaches facilitate the need for a high level of regional acceptance in the context of sustainable development processes and help to better anchor the management aims of protected areas at the local level (see chapters 6 and 11).

The interaction between the concepts and instruments of environmental protection, regional development and spatial planning poses fundamental problems. These concepts, which originate from different sectors, often have an additive character and are insufficiently integrated. This works against the integration of concepts, instruments and corporate bodies which is necessary for sustainable development. This problem frequently occurs with protected area and regional development processes that extend over several regions (see chapters 4 and 6).

The major challenge in many peripheral regions today lies in the linking of close-to-nature tourism, environmental protection and regional development. The use of protected areas for tourism calls for appropriate visitor management and use of natural resources. As an activity of the regional economy, close-to-nature tourism requires specific planning and efficient marketing strategies. The need for greater integration of protected area management, regional management and tourism management is therefore obvious. In this context, sustainable regional management calls for integral process management, providing a framework in which all relevant planning and development instruments could be dealt with and implemented in an integral manner (see chapters 3, 8 and 13).

## Question 3: In Which Areas of Activity (Economic, Social, Environmental, etc.) do Large Protected Areas Contribute to Regional (Rural) Development Processes?

In accordance with the overall generalized aims of the protected areas studied here, three areas of activity can be identified: firstly, protection activities, secondly development activities and thirdly research and educational activities. Protection activities aim primarily at environmental protection (including preservation of habitats) and the preservation of the cultural landscape and its elements (including the preservation of cultural-historical settlement elements). These protection activities coexist with development activities. In most large protected areas these involve the development of close-to-nature tourism, and at a secondary level, the promotion of agricultural or forestry activities. Research and educational activities relate to both natural and cultural themes, i.e. habitats, biodiversity or cultural-historical landscape elements.

Typically, many activities create or are intended to create positive effects not only in one, but in two or even three dimensions of sustainable development. Offers and services such as guided excursions, field observation, study weeks, etc. affect not only the economic dimension (tourism) but also the social dimension (including education, group experience, contact with the local population) and the ecological dimension (including sensitization, explanation of the way nature functions). Or the preservation of a terraced landscape affects the cultural dimension, the ecological dimension (habitat for certain plants and animals) and the economic dimension (attractiveness for tourism). Measures to direct visitors (e.g. educational paths) usually also have an ecological, social and economic significance. Many activities to do with co-ordination and information cannot be attributed to one dimension only. The frequently cross-dimensional character of activities makes it very difficult to calculate the effective social benefits of large protected areas in economic terms and express them in figures. Recording tourist earnings alone does not do justice to the broader significance of the large protected areas for society as a whole.

Among the various dimensions, tourism is quite clearly the most important activity affecting income in the protected areas. Usually, new services or packages of services are created for specific target groups. In order to exploit tourism potential fully, a certain level of pre-existing basic tourism infrastructure is necessary and therefore pre-existing tourist attractions. A large protected area cannot trigger tourism development on its own. It can, however, increase the attractiveness of an area, differentiate the existing tourism services and contribute to a better utilization of existing infrastructure.

In the social dimension, education is the most important activity, with regard to the cultural dimension this role is occupied by the preservation of the cultural landscape and human habitats, and with regard to the ecological dimension it is the preservation of habitat for flora and fauna. Overall, most activities can be integrated into the concept of sustainable development. Individual activities usually take into account criteria from different dimensions and thereby demonstrate a holistic character.

## Question 4: What Kind of Development Schemes or Programmes (National or European) are being Used for this Purpose?

Most large protected areas were created in response to a combination of different incentives from outside the regions and initiatives from within the respective regions. External incentives include state laws, national or sub-national development plans, state and supra-state development programmes or the activities of so-called non-government organizations (NGOs). The aims and intentions of these external elements can vary considerably, ranging from environmental protection through preservation of the cultural landscape to promotion of rural or regional development or a combination of these. In contrast, the motives of initiatives from within the regions are clearly dominated by regional economic development, and the improvement of infrastructure and quality of life for the population.

National and supra-national promotional programmes play a dual role. Firstly, they can spark off the foundation of a large protected area by (co-)financing the preparatory work and therefore gradually including regional actors in the planning and design of the park. Secondly, the promotional programmes finance activities in the protected areas, so that their activities can be extended beyond the mere administration of the large protected area. The activities thereby supported are in a wide range of areas, usually however in the preservation of the cultural landscape and regional development, and they usually include several dimensions of sustainable development. The programmes usually support integrative activities, i.e. those that create added value in several dimensions of sustainable development or aim to do so.

The integrative utilization of regional resources and potential is often in the foreground. Starting with regional raw materials from agriculture and forestry, product chains that bring economic and social benefits are encouraged.

Further instruments used by large protected areas are national and international labels and certificates. At an international level these include the various UNESCO labels such as 'UNESCO World Heritage Site' or 'UNESCO Biosphere Reserve' and at a national level labels such as 'national park' or 'regional nature park'. The functions of such labels are manifold. They have a structuring effect, in that the large protected areas must meet the relevant criteria in order to maintain their claim to the label. At the same time the large protected areas use their labels in order to market their own activities. Thus labels have a similar effect to promotional programmes, especially given that direct or indirect financial support is usually associated with a label. Like the promotional programmes, such labels structure the orientation and activities of large protected areas.

**Question 5: What are the Possible Results/Outcomes of these Activities?**

It is difficult to make quantitatively provable generalizations about the effects of large protected areas as a whole. Some general qualitative statements are however possible, e.g. in the form of identifiable trends.

The direct economic benefits of a protected area for the respective region seem to be relatively low overall. Tourism in particular can benefit directly from a protected area as well as individual related businesses, e.g. in agriculture and forestry, which provide local goods and services. The indirect economic benefits of a large protected area seem to be considerably greater, but it is difficult to calculate these. It is, for example, hard to measure the economic value of a large protected area for regional identity (and that of the urban population), the local population's quality of life, the preservation of the cultural landscape and biodiversity. The central fact that large protected areas can maintain or create direct and indirect economic, social, cultural and ecological benefits should not be forgotten when evaluating the benefits of a large protected area.

Large protected areas provide various services in the social dimension, e.g. via educational facilities, different events, participative bodies (including working groups), the co-ordination of projects. Individuals and groups of actors can exchange ideas and develop projects that lead to co-operative action, local initiatives, new organizations, new institutions, new groupings of actors, innovations and associated activities. It is interesting to observe that as citizen participation increases, regional development tends to be more heavily weighted than environmental protection. In the cultural dimension, large protected areas can contribute to an increased regard for local products and services, their preservation and further development and therefore to the improvement of rural areas, agriculture, the cultural landscape and cultural heritage. With regard to the ecological dimension a wide range of effects can also be observed, e.g. the preservation of habitats or the reintroduction of extinct species.

Integrative concepts for protected areas can contribute particularly to the preservation and further development of the multifunctionality of landscapes and rural regions. It is not easy to determine whether they meet expectations and especially their higher aims, which are usually markedly qualitative in nature. It is difficult to judge whether or to what extent the quality of life of the local population and the future prospects of the individual regions and particularly for the young population can be improved by the creation and maintenance of a large protected area. The higher aims, which are formulated in the respective laws, regulations, charters or guidelines can usually only be reviewed indirectly.

It is possible that the main significance of a large, multifunctional protected area lies neither in the ecological nor in the economic dimension but in the social and cultural dimension of sustainable development. Integrative large protected areas preserve or create social and cultural values as well as the natural and cultural values of a region and thereby contribute to the continued existence of the basic elements that are of central significance for tourism, the economy in general and the quality of life. Thus integrative large protected areas can be seen as important locational factors for the economy and the local population. In this way such protected areas are also suited to regional marketing, tourism marketing and locational marketing.

**Question 6: What Type of Actors are Involved in Development Processes, what Kind of Responsibility do they have and how are these Forms of Local/ Regional Governance Conducted?**

Given the heterogeneity of the protected areas examined here, it is difficult to make generalized statements about the actors involved in the development of large protected areas and the role they play. Nevertheless, some conspicuous trends can be identified which are closely connected to the above observations on the aims and tasks of large protected areas as instruments of regional development and their embeddedness in corresponding concepts and programmes.

The integration of area protection and regional development in the form of large, multifunctional protected areas automatically implies the increasing involvement of a large number of different actors. These include state and economic actors as well as civil society actors from a very broad range of fields who take on responsibility in the context of protected area development. Although the basic conditions can differ widely from case to case and it is almost impossible to compare them directly, nevertheless it is very frequently similar or indeed the same actors who play a central role in the development of protected areas.

The following typical actors or groups of actors can be identified on the basis of the case studies:

The management authorities in the protected areas naturally play a central role in their development. Although they are primarily responsible for administrative matters, their participation in regional development processes is seen as essential, independent of their legal status, financial resources and staffing. Experience has shown that the protected area management authorities usually play a central role in shaping the image of a protected area. Of major significance is the willingness of the management to co-operate with other important actors in the region; thus the management authority of a protected area can play a facilitating role in this respect by motivating different actors and bringing them together (see chapter 10).

Likewise, the co-operation of the affected local communities is essential in all large protected areas. This is true firstly with regard to the participation of the local authorities of all the districts or parts thereof which are included in the protected area; this is also assumed for the participation of neighbouring communes in a broader context (see chapter 6). This refers to the function of the local authorities and municipalities as democratically legitimated representatives of communal interests, through which the viewpoints, expectations and wishes of the population can be expressed and included in the processes of regional development. In some cases the local authorities are also the planning authorities for the territory of the protected area and are thereby included in protected area development (see chapter 11). Municipalities with experience in regional co-operation are also particularly important actors in the development of large protected areas, all the more so when they can provide financial means for their development. In several large protected areas, formal amalgamations of the municipalities form the legal authority in charge of the protected area (see chapter 5). The role of individual mayors should not be underestimated, as they can have a considerable influence on the development of political opinion in a region in their role as promotors (or opponents) of a large protected area.

Diverse higher level state institutions other than the municipalities also play a role in the development of large protected areas. Forestry and agricultural bodies as well as various planning authorities are typical examples of the spectrum of institutions involved. They also highlight the situation of large protected areas in a complex network of relationships including local, regional and national authorities and interests, where they not infrequently play an intermediary role.

A major part is played by the local and regional tourism authorities, as tourism is of central significance for the development of all large protected areas. Today, tourism authorities are frequently organized as private companies (e.g. GmbH or limited companies), and are rarely part of the municipal authorities. In their function as representatives of tourism service providers and in tourism marketing, these play a key role in the development and marketing of a large protected area as a tourist destination. They are responsible for the very important internal marketing which aims to commit the population and actors in a region to its tourist development. In this respect there continue to be marked differences in the level of awareness of and identification with tourism. East Central Europe lags behind in this respect (see Nolte), while this is not a problem in areas with an older tradition of tourism (see chapters 6, 8 and 11).

The involvement of other private economic actors varies considerably from region to region. The inclusion of those directly affected by the development of a protected area is par for the course, e.g. farmers owning land or with usage rights in the protected area, indeed it was often a central prerequisite for the implementation of a protected area concept (see chapters 6 and 8). However the participation of other economic actors is apparently much more difficult. The involvement of such actors can be interesting where the implementation of a brand or a label promises benefits for the marketing of products and services from the protected area. Obviously this is particularly true for tourism, but can also be true for agriculture, trades and crafts, transport or education, as the examples of Donana and Müritz illustrate (see chapters 9 and 10).

The mobilization of civil society actors and the population for the development of protected areas generally presents a major challenge. Actors from the environmental and cultural sectors in particular have significant potential. Not infrequently, the implementation of large protected areas is indeed largely due to their initiative; especially when they are nationally anchored, as illustrated for example by the role of the Austrian Alpine Association (*Österreichischer Alpenverein*) for the Hohe Tauern National Park (see chapter 6). Long-term involvement of the local population is most likely in cases where large protected areas have developed primarily through bottom-up processes and the population was encouraged to participate from the very beginning. Organized processes of citizen participation, such as in the Biosphere Entlebuch, also underline the considerable potential of civil society embeddedness in protected area projects (see chapter 3). Even in cases where large protected areas were ordained by the authorities in a top-down procedure, it was later possible to initiate intensive forms of citizen participation. The main problem in such cases today is permanent and sustainable citizen participation. Future Workshops, Agenda 21 initiatives and other forms of local participation can provide a suitable framework. In comparison to experience in almost all other countries, the mobilization of citizens in East Central Europe continues to demonstrate many deficits. There is obviously a lack of the basic awareness necessary for civil society participation, particularly in the context of the development of large protected areas. The prospects for a speedy improvement of this situation appear to be poor.

The inclusion of individual actors (whether from the state sector, the private economy or civil society) remains inadequate if it takes place without the necessary co-operation of these actors. The establishment and development of partnership-based working structures in the form of networks is therefore a central precondition for the successful direction of development processes in large protected areas, or indeed possibly the central precondition. The essential basis for the activities of such networks is the formulation of shared aims and tasks to which the partners must commit themselves. At the same time these form an excellent basis for the internal and external communication of the protected area. Regional guidelines or development concepts, such as those which are necessary for the implementation of the LEADER programme for example, can fulfil this function. This is also true for the identity-shaping function of a common label in the context of regional marketing (see chapter 10). It is indisputable that all participants must see themselves as equal partners and be in a position to bring their interests to bear in the development processes of a large protected area. Likewise, networks can and must be used as instruments in conflict resolution and for the avoidance of conflicts. The initiation and implementation of specific, commonly defined projects is a tried and tested instrument to deal with conflict-ridden fields of action where shared or at least similar interests overlap, as has been shown in Entlebuch for instance. The preconditions for the stabilization of the networks and their permanent anchoring in the region grow from this basis.

The increasing inclusion of various non-state actors in the development of large protected areas, the growth of partnership-based networks of the actors involved and the testing of participative forms of development planning were identified as typical characteristics in the majority of the case studies presented here. These observations correspond to similar phenomena in other sectors of spatial development. According to Fürst (2003, 46) in the case of large protected areas, these can perhaps be interpreted as 'new forms of spatial planning, the development of which is linked to a general change in responsibilities, jurisdictions and levels of action ...'. These have been discussed for quite some time under the heading of regional governance. The extent to which specific forms of the governance of large protected areas can be subsumed under this heading could not be finally determined on the basis of the case studies presented here. Emerging research, however, indicates that this may be the case.

**Question 7: Are there Clear Differences/Specific Patterns in the Use of Protected Areas as Tools for Regional Development? If this is the Case, where are they to be Found?**

The following generalized broad distinction of the large protected areas discussed in this volume with regard to their function as catalysts for regional development is possible.

Historically, and to a lesser extent in recent times, national parks have primarily served to further environmental protection in the narrower sense. An example is the Swiss National Park. These serve as instruments of regional development or are perceived as such at a secondary level. Where the opposite is the case to an excessive degree, the national park is either not recognized by the IUCN as a Category II, as for instance in parts of the Hohe Tauern National Park (see chapter 6), or its aims and ultimately its territory are pared down in the course of its development. See for instance the problems of intensive ski use in the hinterland of the National Park Vanoise (see chapter 7). Very progressive concepts for protected areas attempt to integrate environmental protection issues directly with regional development. However, the basic conditions, as the Scottish example demonstrates, must be suitable (see chapter 11). The Finnish example clearly shows that national parks, to a greater extent than nature parks, can be utilized as a means of allocating finances in the context of EU regional policy (see chapter 12).

However, the largest group of nature parks (regional parks, regional nature parks, etc.) serves primarily as an instrument for sustainable regional development. The implementation of this concept is perceived more as a challenge for regional policy than for environmental protection policy per se. This is also reflected in the historical development of nature parks as an instrument for the revitalization of marginalized rural areas and for the planning of recreation zones close to agglomerations. The Austrian example shows that although nature parks can be seen primarily as instruments for environmental protection, in reality they are largely used as instruments to promote tourism (see chapter 5).

This differentiation of the two dominant types of protected area in Europe, national park and nature park, can be seen in the example of area protection policy in Switzerland, albeit with features specific to this country (see chapter 13). The catalyst function of the two types, very different in its intentional orientation, is also expressed in the paradigmatic development of these types. This is characterized by a move from an originally more segregative form to a more integrative form in its development from modernism to postmodernism. A certain degree of convergence of these two major protected area concepts can be observed in their 'parallel development' since the 1960s. On the other hand with regard to their IUCN category differentiation and national practices of implementation, very basic differences can be observed. This is manifest in different spheres concerning image, responsibilities and legal regulations for these categories of protected area (see Weixlbaumer 2005, 74).

Biosphere reserves have the most complex qualitative claims. These are, firstly, to have a model character for habitats (see contribution by Hammer) and secondly a bridging function between protection and use. For according to the Seville process, biosphere reserves are defined by an explicit additionality of tools with regard to both nature conservation and regional development. They represent the principle of graded sustainable land-use and aim to have influence beyond their respective biosphere reserve region because of their explicit experimental nature. The historical overlapping with the national park concept which occurs in many cases, clearly shows where the original affinities lie. As the example of Spain shows, where overlapping takes place the internationally recognized label 'biosphere reserve' can also be responsible for the acceptance of national parks as genuine instruments for regional development (see chapter 9). Depending on the dominant regime, an overlapping of protected area structures (in the sense of designated landscapes) can even be productive. This is also evident in the case of Austria, where the relatively 'toothless' landscape protection areas are upgraded particularly with regard to cultural landscape management by being designated nature parks. However, the example of Slovakia shows biosphere reserves can remain bogged down in a segregative understanding of resource administration (see contribution by Nolte). In the post-Seville process, however, the integrative claims of the dynamic and innovative paradigm comes to life. This is very clearly evident in the developments in the biosphere reserve Entlebuch (see chapter 3).

There are, however, considerable differences between the protected areas which can be classified according to these three central protected area types, as already indicated. Thus the spectrum of the nature parks discussed here is very wide:

The Italian example shows that a national park (which by the way is based on a former nature park in the case of Cinque Terre) can primarily be an instrument for the revitalization of a degraded (terraced) cultural landscape and to utilize this in a relatively environmentally friendly way for tourism (see chapter 8). At the other end of the scale, national parks can, as the Finnish example shows, also serve initially mainly as an instrument for the preservation and further development of a wilderness area, and thereby serve as a 'lifebelt' for peripheral regions using a specially developed form of tourism (see chapter 12). In contrast, the East German case of the Müritz National Park Region makes it appear plausible that marketing strategies are equally compatible with national park ideals (see chapter 10).

Because of the considerable degree of heterogeneity even within categories it is not possible to determine which type of protected area is best suited as a tool for sustainable regional development. Nevertheless, the following general conclusions can be drawn: IUCN category II national parks are as a rule more aggressive instruments for nature preservation than category V nature parks. The latter are, however, more aggressive instruments for regional development. The biosphere reserve in accordance with the Seville process, which merely represents a designation (although with a high degree of international authority) is an interface between the two categories. It also aims to go well beyond this as a model region in the true sense of the word.

In spite of regionally varying substrata (such as heritage and cultural artefacts, political regime, planning culture, laws, etc.) which will and should always exist in all their diversity, some of the developments in the heterogenous European area protection policy discussed in this book show that the following shared elements are significant as win-factors:

- Protected area development works best when functioning regional governance and/or government arrangements can be utilized at the time of foundation of the protected area
- When protected areas are planned based on best practices and taking regional and national conditions into account

- That protected areas gain dynamism in their development when they are not afraid to take unconventional paths
- When a close symbiosis is established with the EU regional subsidy structures from LIFE to INTERREG while keeping the aims of the protected areas in view.

Although many of the case studies presented here do not allow a final judgement of the development of the selected protected areas, in the context of the thematic complexes discussed here the question of the model character of the new protected area policy can be answered: The great majority of examples have shown that protected areas are increasingly seen as a multifunctional spatial category and are deliberately used as instruments for regional development. This is true for almost all of the countries discussed here as well as all the types of protected area examined, although there are some marked differences. The experimental nature of many of the development approaches sketched here is characteristic. If therefore it is really possible to speak of a new model for the 21$^{st}$ century in the sense of dynamic and innovative area protection, then this is closely linked to the idea of a 'future workshop' where different forms of specific implementation and development of the model can be tested in practice.

It is remarkable that this workshop situation is true of countries with very different starting points: It is characteristic for long-established systems of area protection policy, e.g. in Germany or Austria, where protected areas, especially the nature parks and biosphere reserves, are the vehicles of the new paradigm and are being tested experimentally as such. This is also true of countries without a long-established tradition of large protected areas. The example of Scotland represents a new beginning at the start of the 21$^{st}$ century, which could be exemplary for many others. Finally, this could be true for the difficult situation in the countries of East Central Europe. Although the paradigm of static, conservation-oriented area protection continues to dominate here, there is some hope that the experiences with new forms of area protection in other European countries could affect the situation in countries such as Slovakia.

In the context of the many and varied case study analyses there is no question that with the paradigm of dynamic and innovative area protection a new course has been set, the logical result of which is the increasing integration of environmental conservation and regional development. This model can work in the face of the challenges of the 21$^{st}$ century. Nevertheless, it remains dependent on political support, financial resources and the willingness of its actors to work in partnerships in order to become established as everyday practice.


## References

Fürst, D. (2003), Regional Governance. In: Benz, A. (Hrsg.): *Governance – Regieren in komplexen Regelsystemen. Eine Einführung.* (Opladen), 45–64.
Habermas, J. (1987), *Die Neue Unübersichtlichkeit.* (Frankfurt am Main).

Mönnecke, M. and Wasem, K. (2005), *Sportaktivitäten im Einklang mit Natur und Landschaft. Handlungsorientierte Lösungen für die Praxis*. (Forschungsstelle für Freizeit, Tourismus und Landschaft. Rapperswil) (Download: http://www.ftl.hsr. ch).

Weixlbaumer, N. (2005), 'Naturparke' – Sensible Instrumente nachhaltiger Landschaftsentwicklung. Eine Gegenüberstellung der Gebietsschutzpolitik Österreichs und Kanadas. In: *Mitteilungen der Österreichischen Geographischen Gesellschaft*, 147. Jg., 67–100.

Yin, R. K. (2003), *Case Study Research. Design and Methods*. Third Edition. (Thousand Oaks, London, New Delhi).

# Index